ETHEREUM FOR BEGINNERS

The Complete Guide on How Ethereum Works

(The Blueprint on How to Buy, Sell and Make Money With Ethereum)

Michael Gates

Published by Tomas Edwards

Michael Gates

Ethereum for Beginners: The Complete Guide on How Ethereum Works (The Blueprint on How to Buy, Sell and Make Money With Ethereum)

ISBN 978-1-990373-64-0

All rights reserved. No part of this guide may be reproduced in any form without permission in writing from the publisher except in the case of brief quotations embodied in critical articles or reviews.

Legal & Disclaimer

The information contained in this book is not designed to replace or take the place of any form of medicine or professional medical advice. The information in this book has been provided for educational and entertainment purposes only.

The information contained in this book has been compiled from sources deemed reliable, and it is accurate to the best of the Author's knowledge; however, the Author cannot guarantee its accuracy and validity and cannot be held liable for any errors or omissions. Changes are periodically made to this book. You must consult your doctor or get professional medical advice before using any of the suggested remedies, techniques, or information in this book.

Upon using the information contained in this book, you agree to hold harmless the Author from and against any damages, costs, and expenses, including any legal fees potentially resulting from the application of any of the information provided by this guide. This disclaimer applies to any damages or injury caused by the use and application, whether directly or indirectly, of any advice or information presented, whether for breach of contract, tort, negligence, personal injury, criminal intent, or under any other cause of action.

You agree to accept all risks of using the information presented inside this book. You need to consult a professional medical practitioner in order to ensure you are both able and healthy enough to participate in this program.

Table of Contents

Introduction

The following chapters will ensure that you are ready for what's ahead, starting with discussing the fundamentals of the Ethereum blockchain and what sets it apart from others of its ilk. From there, you will learn about some of the changes that lie ahead when it comes to how smart contracts will remake infrastructure of all types. You will then learn all about trading ether, the Ethereum currency, as well as all the other cryptocurrencies that have been created on the Ethereum blockchain.

From there you will learn about mining Ethereum, why it is essential, how you can make a profit from doing so and how to build your own mining machine on the cheap. Next, you will learn the basics behind creating a decentralized application using Ethereum and programming smart contracts as well. Finally, you will learn where Ethereum is headed, both in the short and in the long-term.

There are plenty of books on this subject on the market, thanks again for choosing this one! Every effort was made to ensure it is full of as much useful information as possible, please enjoy!

Chapter 1: What is Ethereum?

The Ethereum platform has been in existence for a couple of years but has just recently begun the forward march to success, going into partnership with Microsoft to release software, present the BaaS toolkit and so much more besides. It is more often referred to as the "brainchild" of Vitalik Buterin, a Russian-Canadian prodigy in terms of computer programming. When he was just 19 years old, he was a real enthusiast of sociology, politics, economics, epistemology, cryptology and information theory and, in 2013, he released the first white paper on his idea.

He referred to the invention as a decentralized application platform and a next-generation cryptocurrency, explaining that, rather than the complicated protocols designed for specific needs for specific industries, he wanted Ethereum to be as generalized as it possibly could be. This would enable people to create their own specialized applications on Ethereum for just about any purpose they wanted.

He also argues that, although Bitcoin was perfect for the initial purposes of storing value and transferring it, it isn't quite so effective as a universal platform that can be used to build any number of decentralized applications, or DApps. As a result, he opted to produce an alternative, a protocol that would allow people to use smart contracts, which are applications that run exactly as they are programmed to do, without any fraud, censorship, downtime or interference from third parties. He also wanted this to be for both financial purposes and non-financial.

In 2014 a yellow paper was released by Dr. Gavin Wood, a colleague of Buterin's. This paper was a draft of 32 pages detailing the computations and the substantiations that described the EVM or the Ethereum Virtual Machine. The paper details Ethereum as being a sophisticated and "specialized version of a cryptographically secure and transaction-based state machine". This, he said, had the potential to ensure that transactions were trustworthy, something that would not be possible otherwise because of geographical separation, and the expense, unwillingness, incompetence, incompatibility and the inconvenience, not to mention the corruption of the existing legal systems. To fix all of this, Ethereum produced a system that was a "disinterested algorithmic interpreter", something that could track every step and show how any judgment or particular state came about.

Ethereum was officially announced on January 25, 2014, and a crowdfund was set up. By August, the required donations had been received from all over the world and were followed by the establishment of the Ethereum Foundation in Switzerland. This was set up as a non-profit organization that was designed to encourage developers to come up with the next generation of DApps using Ethereum as a platform.

The first live Ethereum blockchain was launched in July 2015, with the Ether, the unit of currency, being crowdsold up front in a public round that lasted 42 days. The Ether or ETH token could be exchanged into Bitcoin and, rather than being described as a digital currency, rather it is seen as a kind of fuel for the Ethereum platform. To begin with, it was never intended that ETH would ever be used as an asset, a share or a currency. It is, however, a form of payment that the platform clients make for requested operations to be performed. In a more basic explanation, it is an incentive that ensures all developers produce quality code because bad code costs more to run and that the network stays in a healthy state – users are given compensation for any resource that they contribute.

According to Buterin and Wood, Ether is not designed to be a competitor to Bitcoin; instead, they are complementary systems, each with their own digital ecosystem place. Ethereum provides investors with a way of producing their own unique currency that is pegged to the Bitcoin value, so long as all interested parties trust the price source.

As such, Ether was not initially released for market speculation but for developers to build their Ethereum-based apps or to get involved in smart contracts. During the presale in 2014, the issuance rate and the total supply were defined. At that point, 60 mln Ether were given to the contributors, with 5 Ether per block (this equates to about 15 to 17 seconds) as a reward for the miner of that block. Also during that presale, an agreement was reached that a limit of 18 mln Ether would be issued each year, just 25% of the initial Ether supply.

To limit smart contracts that may take time to run, Ethereum uses "gas", a mechanism that measures the computation effort that evaluates each transaction. This value is the internal pricing for the running of a contract or a transaction in Ethereum.

What Can Ethereum Do?

For now, the Ethereum platform gives us an opportunity to create contracts, smart contracts that hold on to money contributed until a specific date or goal is attained. And when that date or goal is reached, the funds held will either be given back to the contributors or, if successful, are doled out among the owners and developers of the project. This allows crowdsales, crowdfunding and auctions to be held without interference or any need for a third-party to be involved and with the dispersal of the outcome being pretty predictable.

Another more sophisticated possibility is that smart contracts may be used to run entire organizations. Rather than having to hire people to run board meetings, handle accounts and do a whole heap of paperwork, it can all be left on one simple contract built on Ethereum. Because the "robot" can't be influenced by any outside factors, it will execute exactly what it has been programmed to do and the collection and distribution of proposals from members of the organization would be submitted in a voting procedure that is transparent.

Ethereum is being presented as a tool that has a diverse range of different application possibilities, allowing developers to build, according to the official website:

Virtual organizations that allow issues to be voted on by members

Transparent associations that are based on votes by shareholders

A country with a constitution that cannot be changed

A better democracy

Some of the options that can be built on Ethereum, either using the Blockchain as a Service toolkit on Azure or directly, include:

Invoices that will pay themselves when shipments are successfully delivered

Share certificates that will send dividends automatically when profits hit a specific level

Ethereum is, in short, seen as one of the biggest steps towards the Internet of Things and a world that is decentralized, with just one limit – our own imaginations.

Chapter 2: The Future of Ethereum

Just like every other piece of technology, you are not going to know the future and what it will do. Ethereum is a platform that will offer you more than bitcoin, and that makes it to where it will have the possibility of outlasting bitcoin.

Since Ethereum is continuously changing, it will run the risk of crashing and burning or continuing on as it has been. The evolution as you have seen is not always a bad thing, but it is something that can stand in the way of your investing with Ethereum. With the servers continually going down, you may find that you do not want to invest in something that is so unstable and is making it to where you are unaware if you will be able to do what you have to do on a day to day basis.

You will have to watch the news to ensure that you can tell what the market is doing so that you can figure out what is going on with the platform. Ethereum will be similar to investing in the stock market. It will have its ups and downs, and if you are unprepared for one of the falls, you will lose your coins.

If you invest with Ethereum, you will hold onto the hope that it will stick around so that you can continue to trade on the platform as well as mine coins.

Since blockchain is changing so many sectors in the world, is it possible that the blockchain applications will stick around longer than anyone can predict? Or, is the government going to find some way to step in and begin to govern them or get rid of them all together. Cryptocurrency is already subjected to tax therefore it is already monitored by the government. However, how far is the government going to be willing to go to tell you what you can and cannot do with your Ethereum cryptocurrency?

Unfortunately, no one can tell the future. Therefore no one will know what the future holds for Ethereum and other blockchain applications. One-minute Ethereum could be here, and the next minute it could be gone because of the government or because of traditional banking systems. There is no telling what will happen with Ethereum sadly, so you cannot say that it will stick around. However, you can look at how good it is doing and make an educated guess that it will stick around.

If you look at some of the cryptocurrencies that have already vanished, they vanished overnight. However, how do does anyone know what is to happen with the ones that appear to be stable?

If you were able to tell the future, the chances are that you will be able to see that Ethereum will still be around because of how much it has to offer its users. You have not seen a lot about Ethereum yet in this book, but you will see that Ethereum is a vital asset to those who are investing with digital currencies.

Maybe sometime in the future, Ethereum will open themselves up to other cryptocurrencies because they will want to make a more efficient platform and opening themselves up would be something that would not only help the Ethereum users but would help draw in users who use bitcoin and even litecoin.

There have not been any announcements as to what Ethereum has to offer its users lately, but there is probably going to be several projects in the works for Ethereum that they do not want the public to know about yet because they are wanting to ensure that the coding is just right before they release the projects for beta testing or any other further testing that will be required by the users.

Chapter 3: Understanding Ethereum

In this chapter, you will learn and understand the basic aspects of Ethereum. This information will help you get a deeper understanding of this cryptocurrency. You will start by learning about blockchains, how they operate, and how Ethereum itself works. By the time you are through with this chapter, you should be able to understand the foundation of this cryptocurrency.

To understand Ethereum, you must first understand what a blockchain is.

Blockchains

A blockchain is a public ledger of digital transactions. The name comes from joining the words "block" and "chain." A block refers to a digital transaction. When multiple blocks are joined together, they form a chain. In such a system, the transactions are added in a chronological order.

We can also say that a blockchain is a database that is always expanding and contains a large amount of data and information. One of the primary qualities of the blockchain is that when people store data in it, deleting or modifying that data becomes impossible. Data in the blockchain stays there forever. You should also know that no single individual maintains the database and no organization manages it. It is a collective effort by different users who support the database from various parts of the world. As a security measure, they all have different copies of the database.

Here is how to understand how the database works. Think about a scene where you and ten friends each sitting with an empty basket. When one person adds a fruit into their basket, they have to announce it to everybody else in the group. Everyone takes note of what the announcer has said. They note the changes down until they all fill their baskets. After everyone has filled their basket, they close that process and move on to the next one.

Now, think of the fruits as mathematical puzzles. When they solve these puzzles, they add them into the system and guarantee that no one can ever change the contents of that system. The person who can solve the problem the fastest within the system gets paid a certain sum in the form of a cryptocurrency. The solution to the mathematical puzzle is what we call a block.

After that, another mathematical equation comes and the solved one is locked in a folder. The process goes on and on and users keep updating what they are doing on the network. These transaction blocks can be stored in folders called chains hence the name "blockchain."

A blockchain stores different types of data. The value of the blockchain depends on the type of data it contains. For example, a Bitcoin blockchain is one that has financial data that is comparable to dollars. However, since these currencies are not tangible, they are referred to as cryptocurrencies. Bitcoin is very similar to dollars, but as you will soon find out, Ethereum is not.

The blockchain is unique because when someone inserts data into the system, they do not remain with it. Everything in the system is a copy and any user can access their copy. Think of it this way. If a blogger puts an article on a website, the article can be read by millions of people all over the world. Everyone can download or access a copy of the webpage on their browser. This is how the blockchain works. When a user performs a transaction, the copy of the transaction can be accessed by virtually everyone on the system.

Understanding Ethereum

Ethereum is a free-to-use platform that is based on blockchain technology and facilitates users to install integrated applications. Think of Ethereum as a giant, slow computer that produces a currency known as Ether. When put side by side with a conventional machine, it is a 100 times more time-consuming. Though it is as slow as the first cell phones that people used back in the 1990's, Ethereum is actually quite costly and only undertakes the most minimal operations.

You may be wondering why people are making such a fuss about Ethereum yet the platform is so slow. Well, to answer that question, you must first understand how this system works.

How Ethereum Works

Ethereum is based on the blockchain model. There have to be very many users in the network for it to work. All the computers in the network are referred to as "nodes," with each serving a unique purpose. Each node runs the EVM (Ethereum Virtual Machine), which is an operating system that uses a programming language known as Smart Contracts.

People who want to start a process on this network have to pay. The node operators pay with a cryptocurrency rather than money in dollars. The currency that they use is called "Ether." It works in the same way as Bitcoin but is specific to the Smart Contracts or processes that run on Ethereum.

The incredible thing is that Ethereum views human beings and Smart Contracts in the same way. For example, if you are on the network, you can receive Ether currency just like the Smart Contracts. However, the difference is that Smart Contracts perform different processes using the computer language while human beings may not have the ability to carry out those operations as fast as required.

All this may sound complicated but here is an example to elaborate it further. For example, you and your friend have a bet. You say that team A will win a match while your friend bets on team B. The loser is supposed to give the winner $50. How are you two going to ensure that you both adhere to the rules of the bet and no one will go back on their word?

There are three ways to guarantee the bet:

Trust - If both of you trust each other, it will be easy to fulfill the promise.

Get a third party - A third party will ensure that the winner gets the money. However, this is not 100 percent safe because the third party can steal the money.

Sign a binding agreement - A legal contract will guarantee that the loser pays the winner the money.

This is how the Smart Contracts help to secure transactions on Ethereum. It is similar to the second option of the third party, except there is a code that guarantees the contract. The Ethereum network enables people to have such bets and agreements and ensures that the winner gets the amount promised.

This information leads us back to the point where we said that no one can change the data in the blockchain. In short, no one can delete or edit the Smart Contract once it has been generated. Bringing it all back in, once the Smart Contract is in writing, the computer stores the transaction in a blockchain.

That explains how the Ethereum blockchain works. As long as there is a Smart Contract, the users and nodes on the network will put it into action no matter what happens.

This chapter has explained some of the terms that you will be coming across in the other chapters of the book. Before proceeding to the next section, ensure that you understand what Ethereum, Ether, blockchain and Smart Contracts are. If you do not, go back to the chapter and familiarize yourself.

In the next chapter, you will learn how Ethereum is different from Bitcoin.

Chapter 4: Understanding Ethereum Basics

Now that you have been provided with an idea of the climate surrounding Ethereum's emergence and what makes Ethereum distinct from other types of cryptocurrencies that are on the market, we will now begin to work towards developing your intricate understanding of what Ethereum is and how it functions. After reading this chapter, you will have a working definition of all the major aspects of Ethereum. You will also be given information regarding multiple types of currency that exist within Ethereum, and the controversy surrounding these various currency types. This way, you will know exactly what type of currency to purchase if and when you decide to invest in Ethereum.

The Ether Currency

Perhaps the most rudimentary term integral to the Ethereum network is its currency, known as ether. Without owning any ether, you will be unable to operate anything on Ethereum's platform. In other words, ether is what an Ethereum user will purchase if he or she is looking to participate within the Ethereum network. You can compare this to having to buy a subway pass prior to traveling to various places in New York City. You want to send yourself from point A to point B, but you cannot do this without first subscribing to the subway system in the form of payment. Unlike a train network, the cost of 1 ether is going to depend on the demand for the currency itself. Right now, 1 ether is going to cost you around $225. If you're interested in purchasing ether, you can either find a physical exchange place where ether is sold, or you can mine the currency itself. We will get into more detail about mining in a moment. On the other hand, some websites where you can purchase ether with US dollars include Coinbase, Gemini, or Kraken. Please keep in mind that these sites are going to require that you verify your information via identification documents prior to being given access to the ether coin. Still, it might make sense to at least check out these websites prior to purchasing any ether at all.

The Notion of Mining

Similar to when the people of the Wild West mined for gold, ether also must be digitally mined. This is not done with a shovel or chisel, but rather through cryptographic puzzles that must be solved. As was stated previously, a blockchain network is comprised of multiple computers who all collectively use their authority to verify transactions. The people behind these computers are known as miners. An additional duty that the miners have is to extract ether through cryptographic puzzles. Not only do these puzzles ensure that new ether is being created; it also helps to secure Ethereum network from hackers through encryption upkeep. A primary reason why mining is necessary is so that double spending is more difficult. Because miners are mathematically verifying that the transactions that are taking place are true, it's more difficult for anyone using the network to illegally spend the same ether twice.

Hash Functions

Also known as a hash algorithm, this is mathematical equation that the miners must solve when attempting to upload new ether to the Ethereum network or upload a block of transactions to the blockchain. Hash functions are integral to any blockchain network, but the hash equations themselves are going to differ depending on the blockchain network you're using. On average, hash functions take about ten minutes to solve. The goal of solving a hash function is to produce a nonce number. A nonce number can be defined as a predefined number that the miner is looking for while solving the hash function. In exchange for utilizing a lot of their own computer space and solving these hash functions as quickly as possible, a miner is typically going to be paid in some fashion. Within Ethereum, this means that the miners receive ether in exchange for their hard work supporting the system as a whole.

Smart Contracts

An entire chapter of this book is dedicated to the ins and outs of Smart Contracts; however, Smart Contracts are so integral to Ethereum operation that they need to at least be mentioned here. A Smart Contract can be best defined as a digital agreement between two parties. Similar to a financial transaction, a Smart Contract allows two parties to involve themselves in a transaction, without necessarily relying on one another for the deal to complete itself. For example, let's say that Bob is looking to sell his home. Sally decides that she wants to buy Bob's home. If you've ever purchased a home in the past, then you already know that the paperwork involved with this task can be painstakingly boring. A Smart Contract essentially fast-tracks this process for you. In other words, in the physical world, all Sally and Bob would have to do is agree upon a date for the Contract to begin.

Once that date arrives, the Smart Contract would automatically engage, at which point it would notify both Sally and Bob of its completion. Again, we will go much deeper into Smart Contracts in a subsequent chapter.

Dapps

Dapps is an abbreviation that stands for decentralized application. The best way to describe a Dapps is to first think about how a typical application works. An application that is open on your phone, for example, can only operate when it's running on a central server. In other words, only a few computers are able to run the application. Contrastingly, a Dapps application is designed to operate on only decentralized computers. Therefore, these applications are designed to work on the computers that the miners use along the Ethereum blockchain.

Swarms and Whispers

Within Ethereum, the goal is to have the network operate as a global computer.

Ideally, any decentralized computer that's been programmed to run Ethereum should be able to reach it. While the blockchain technology behind Ethereum largely enables any computer to talk to one another, the Smart Contracts that are integral to the Ethereum application require excess storage for two crucial reasons. Firstly, the Smart Contracts need to remember the parameters of the Contracts that are being signed. If there's limited storage space, memory is limited. Secondly, in order to communicate properly across the Dapps, each decentralized computer needs to have enough bandwidth to do so.

To compensate for these needs that the Smart Contracts have, the concepts of swarms and whispers were produced. The swarm acts as the storage that allows for greater memory, while the whispers provide the greater bandwidth that the Smart Contracts require. The image below should provide greater clarity for these concepts:

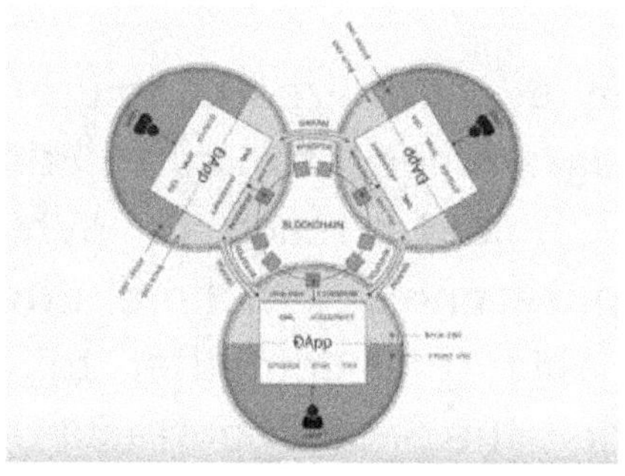

The three circles in this image represent three separate decentralized computers. As you can see, both the swarms and the whispers have been strategically placed to connect all of the Dapps to one another. You can also see that both the swarms and the whispers are placed in what's known as the "back end" of the Dapps. The back end is the decentralized aspect of the Dapps. In other words, a Dapps computer is programmed so that it's essentially cut into half. The front end of the computer houses the centralized Ethereum network. The CSS, HTML, and QTQUICK code all contain the information that is essential for the Ethereum website to run, while the back end of the computer communicates with the front end of the

computer through the points where the
swarms and the whispers interact.

Chapter 5: Technology Engine supporting Ethereum

Just like the Internet, Ethereum isn't just 1 thing, however, is instead a sum of many components. This chapter investigates the

building blocks behind the Ethereum. Now, let's get started.

Ethereum Technology heap along with also the big imageThe Ethereum system Includes a non-exhaustive List of elements with

cryptographic tokens, the speech system, miners or system of validators, consensus algorithm, a blockchain ledger, the Ethereum,

Virtual Machine, programming languages, scripts and complicated economic structures.

The diagram below summarizes the Ethereum Technology heap:

Mist browser

Decentralized programs (Dapps)

Swarm (storage)

Whisper (Messaging)

Oracles

Hardware Customers

Ethereum Supporting ProtocolsAt the bottom of the Ethereum technologies is The hardware system. For your own hardware system to

encourage Ethereum, then you'll need fast processing power together with the following specifications:

· CPU. When choosing the CPU, then you need to opt for the bare minimum to manage high computations.

· RAM. You ought to have a computer using a minimum RAM of 4GB.

· GPU. Ethereum computations are carried out within the GPU. Thus, a strong GPU that's high computational capabilities are your

best bet.

· Hard drive. For the hard disk drive, using an SSD is critical.

More details about the hardware system Specifications will be provided in a later chapter.

The second layer of the Ethereum stack Technology comprises of software and their accompanying protocols. These rules support the

growth and development of Ethereum by complimenting the network to generate the parts run more efficiently. The most Frequent

protocols in this layer are:

· Whisper. Whisper is a communications protocol and toolset that allows programs constructed on the Ethereum protocol stack to

talk to one another. It unites all the aspects of the distributed hash table and point-to-point communications system to permit programs on the Ethereum system to communicate together.

· Swarm or Swarm Hash. Swarm or Swarm Hash is a peer-to-peer file sharing protocol that is intended to effectively store and

retrieve information for use in Ethereum programs and contracts. The simplest analogy to draw with Swarm would be that it is

basically the BitTorrent to get Ethereum.

· Oracle. Ideally, we need to have a construct that communicates the outside realities to the wise contracts. Back in Ethereum,

the constructs that communicate with the outside realities are known as oracles. Though a number of the jobs create personal

oracle systems, there were attempts to develop a single system to help confirm inputs to numerous Blockchains.

The next layer of this Ethereum technology Heap is Dapps. Dapps can be shaped from one DAO or a series of DAOs that work in unison

to create an application. This may result in something similar to Google Chrome or Microsoft Outlook. Such apps may be made to

achieve a certain functionality

In the surface or the application layer of The Ethereum heap technologies is Mist. Mist provides users with the capacity to

explore the programs and offerings that use the Ethereum protocol. Designed as a dispersed program detection instrument, Mist

serves as a wallet for the wise contracts that allows GUI-allowing users to place their transaction fees and handle customized

tokens dynamically.

Scripting language A scripting language is simply a programming Language which supports the growth of scripts (applications that have been created for run-time environments and to minimize the need for human intervention). Due to this property, scripting

languages are be best exploited for experimentation and rapid prototyping.

The Bitcoin system has a primary scripting Language, and there is justification for it. From the start, Bitcoin's programmers have

always prioritized the capacity to "push" motion of Bitcoin via the Bitcoin Blockchain rather than other applications. While talks

are at an advanced stage to integrate a more powerful scripting language which promotes program growth, the Bitcoin community has

mostly remained reluctant to embrace new functionalities.

Actually, the neighborhood has given credence to Censorship resistance and community security at the expense of a more powerful

scripting language that could enhance the development of apps. Ethereum, on the other hand, aims to function as "Turing-complete"

system. This implies that, if any user has infinite resources like memory, computational power, and storage, then the infinite

"loops" could be executed.

In other words, both the logic and Functionality which may be integrated in Ethereum transactions is only restricted by the access

to the protocol currency. But this feature comes at the price of enhanced security. While powerful scripting enhances greater

features, the further tools also create new security challenges for your user.

Ethereum TransactionsThe most noteworthy difference between Bitcoin

And Ethereum transactions is that Ethereum blocks contain both

a trade list and the latest condition of the ledger. This helps to handle two Major types of accounts:

· Externally owned balances (EOAs). EOAs interact with and create normal updates around the Ethereum blockchain.

· Contracts. Contracts programmatically execute only when they get directions, which are in the form of a trade, from an EOA. The

contracts can either push or pull funds and may also request these activities from other contracts while calling on the contract

code to carry out dynamic actions.

Ethereum doesn't utilize the transaction inputs or Outputs, which further in the Unspent Transaction Outputs (UTXO) model of the

Bitcoin that has been popularized by Satoshi Nakamoto. From the Bitcoin's model, each newly minted Bitcoin becomes an UTXO with

the owner who keeps the right to absorb that Bitcoin later on.

Throughout any Bitcoin transaction, the UTXO Becomes the input that's spent in the trade. When these Bitcoins are consumed or

pushed into a different Bitcoin consumer, a brand new UTXO is generated. By comparison, Ethereum stores the present condition of

its community, such as a full collection of their accounts and their related accounts' accounts.

Rather than confirming that the UTXO spent in The trade are valid, Ethereum network determines whether the sender gets a

Sufficient account balance, just like a bank verifying if a specified check Can be cleared. This feature becomes vital when the transactions include Pictures as recipients. If the transaction recipient is also contracted, then That contract's piece of code

will also be implemented, changing both the state of That contract and tripping other contracts to implement codes as well.

The Ethereum Blockchain

Exactly like Bitcoin, the Ethereum ecosystem also Operates global trade of ledgers that achieves remote and distributed validation

using a Proof-of-Work (PoW) protocol. The PoW is a consensus mechanism where participants invest a substantial quantity of energy

in identifying particular pieces of data which can be verified by the broader network.

This information is utilized to Create the Blocks, or specific finite volumes of trade data, which serve as a benchmark for other

network participants. The consequent Blockchain can offer a history of the Ethereum system at each of those periods, therefore

creating a shared reality that relates to all transactions.

The cubes from both Bitcoin and Ethereum have been Similar in that they contain details such as the block amount (which specifies

the number of cubes that have passed since the first block) and also the problem (a metric which specifies how hard it is to

calculate the work required to generate a block). On the Bitcoin ecosystem, the trade script is stateless meaning that there is no

condition prior to the execution of the script. Any update to this state is not stored after its execution.

Contracts on the Ethereum system are Considered stateful in the feeling that they're conscious of their previous information that

was stored on the network and when they are instructed to via smart contracts, they can then do it. In other words, these

contracts can be given a block of information and run all of the trades to verify any mathematical figure representing the machine

state at the instant in time. In the event the network nodes can validate the information, then they will take the block for

addition on the Blockchain.

Block dimensions, Blockchain dimensions and Block TimesOn the Bitcoin blockchain, the cubes are Limited to 1 MB in size. This not

only creates a caveat on the quantity of trades that may be processed in a given time, but it has also proven to be a significant

point of debate within the Bitcoin community regarding whether to stick to it or to increase the size.

Ethereum on the other hand has no such limitation On the block dimensions. This is necessary since Ethereum executes both the

scripts and the contracts. Capping the block size would not only impede the concept of "Turing-completeness" but would also

restrict the volume of storage that a contract could use to execute.

Rather than restricting the magnitude of its cubes, Ethereum uses a mechanism which produces contracts more costly to implement if

they are bigger in proportion.

What about the Blockchain size?

The procedure for allocating the Blockchain dimensions Is akin to that of the Bitcoin network. The larger the amount of

transactions executed on Ethereum, the more data each of the peers in the machine will want to store. This is because of the fact

that the necessity to keep track and store all these trades requires funds from the community of computers which are running the

Blockchain. As of May 2017, the size of the Ethereum blockchain has risen to roughly 11 GB.

While this is still smaller than the Bitcoin Network's blockchain size of approximately 69 GB, it's worth noting that Bitcoin has

been around for a longer period when compared with Ethereum. Assuming a mean growth rate of about 1 GB a month, the Ethereum's blockchain is still growing at snail's pace in contrast to Bitcoin's. However, Ethereum has gained considerable grip because its

genesis block. As the system becomes more popular, its own monthly growth rate may increase.

When it comes to Block times, Ethereum is Faster in comparison to Bitcoin. The block period has been put from 14 to 15 minutes

using a Ghost protocol while that of Bitcoin is 10 minutes. Therefore, Ethereum transactions are faster when compared to Bitcoin

trades.

Consensus AlgorithmsFor almost any decentralized computing system to Function correctly, there needs to be some sort of mechanism

by which the entire network may come to an agreement with its state, or how its token supply is divided among the nodes that are

registered on the network.

The Bitcoin system uses what's commonly Known as "Nakamoto consensus." Indeed the revolutionary innovation supporting the Bitcoin

technology, Nakamoto's invention solved a longstanding complex computer science problem that's often known as the Byzantine Fault

Tolerance or Byzantine Generals' Problem.

It's conceptualized on the thought that one Person can not trust another person who has the possible motivation to lie and one

can't entirely trust the ethics of any given communication if it moves through a third party. Bitcoin solves this job by making a

chain of proof of work. The miners on the Blockchain spend substantial amounts of electricity to fix a complicated cryptographic

algorithm in a bid to receive rewards when they locate another block in the Blockchain.

Considering that the following block always paths the Previous block (indicating that you begin the algorithm from the tip of the

block), then Bitcoin miners can hurry to verify that the cube is valid and proceed to find the next block so that they maintain a

reward. Ideally, mining is incentive compatibility, as well as the immutable nature of the admissions at the Blockchain provides a

solution to the Byzantine Generals' problem.

Even though programs are underway to migrate the Bitcoin network to a different protocol within the next few decades, at the time

of writing, Ethereum still utilizes a comparable PoW protocol that's called Ethash. So, just how is Ethash different from PoW?

In Bitcoin's PoW, the miner assembles a candidate Block that is full of transactions. Once the candidate block is set up, the

miner computes the hash function--the algorithm that determines the integrity of information--to ascertain if it could fit in the

current target Blockchain. If the hash algorithm can't match in the current Blockchain, it will automatically upgrade the

Blockchain by incorporating the transactions to it.

On the other hand, the Ethash uses different Cryptographic primitive because of its hashing function--SHA-3 or rather SHA-256 to

confirm the trades on the Blockchain. While the differences are nuanced, Ethash has been designed to make Ethereum resistant to

the high profile mining chips which are now dominating the Bitcoin industry.

It also makes it more available to "light" Client implementations that empower users to use Ethereum without needing to first to

download the Ethereum blockchain for their apparatus. These days, the majority of all Bitcoin mining is done in data centers that

are largely VC-backed companies that handle the manufacturing cycle of the gear and collaborative collections of individual miners

that are generally known as mining pools.

To mitigate this consolidation, Ethereum Mining was set up so that it can only be achieved with strong GPUs. The system is

permission-less, meaning that any node that buys a GPU and decides to conduct an Ethereum client can start processing

transactions. However, if the intended switch to the new "Proof-of-Stake" consensus algorithm happens, mining may no longer be required in the near future.

Ethereum MiningEthereum mining is the process of using your Pc to make blocks that are utilized to verify and process the

transactions that occur in Ethereum systems. The blocks that your computer produces contain information from other preceding

cubes. The practice of producing these blocks is tedious and demands a lot of computer processing power.

Without an incentive, no consumer will bother Wasting their time producing blocks in the Blockchain. Therefore, as an incentive,

any person who generates a block successfully is well rewarded.

Solo or swimming mining? Ethereum mining is very similar to actual mining If you think about each GPU to be the consumer doing

that exploration.

Solo mining is if you mine the Ethers by yourself. The advantage of solo mining is that whatever you find is entirely yours.

Unfortunately, in the event that you only have a couple miners, it can take you a great deal of time to discover that Ether.

Anyway, the frequency of getting the Ether may also vary considerably.

You can have a week in which you hit on the Ether Thrice, but then nothing for an entire month. Evidently, if you've got many

miners--or GPUs for that matter--then the outcome will be more stable. But any threshold that has a mining rate of less than 1GH/s

isn't advisable if you're aiming for profitability. If you are not bothered about the changes in that you locate Ether, then solo

mining might be a good alternative for a mining speed above 100Mh/s as you don't have to pay fees to anyone.

With pool mining, many miners--or even GPUs--combine Forces to try and locate Ethers. The Ethers found are then shared equally

among the miners, although some pools choose to distribute the benefits in the kind of ratios based on numerous factors.

Nonetheless, in pool mining, you are required to pay a small fee (generally less than one % of the reward) into the pool operators

to keeping up the mining service. The upside of pool mining is that you are always going to have consistent payout and will

consequently make more income.

Ethereum Basic Hardware and SoftwareMinimally, you'll need a computer (using a quick Processing power), the application for mining

the Ether and high-speed Internet connection. Below is a list of specifications for your hardware you will need:

· CPU. When choosing the CPU, then you should go for the bare minimum. I really don't advise purchasing the cheapest pc. This is

because extra processing power will be required to perform computations.

· RAM. You ought to have a computer using a bare minimum of 4GB. If you're choosing solo mining, it's important to get as much RAM

potential as you can to boost your chances of success.

· GPU. Comparatively speaking, GPU-mining in Ethereum is considerably faster compared to CPU-mining. Currently, CPU-mining is no

longer profitable or rewarding. Even entry-level GPU-mining is about 200 times faster than CPU-mining. Presently, RX480s are among

the very popular GPUs which you can locate for Ethereum mining cards. Just ensure it has sufficient processing capacities.

· Hard disk. For the hard disk, using an SSD is critical. While SSDs can be more expensive, you may only need around 16GB, that

will cost you fewer bucks.

What about the Program?

Each mining software which you find in the Market has evolved through the years. But some have developed over others. The main

contenders for Ethereum mining are Geth, MinerGate, Claymore and Genoil.

GethGeth is the first software from the Ethereum Foundation team. If you want to solo mine, then Geth will be your best option. It

is simple and straightforward to use. It can also help you create your wallet. If you want that the GUI option, it is possible to

attempt Mist/Ethereum Wallet.

MinerGateMinerGate isn't the best bet if you're Planning to have committed mining grills. But if you would like to mine on an

present computer as a hobby, then it's perfect. While it takes a fee from your mining process, its own GUI is fast and easy. It

has some challenges that could motivate you to mine if you are a newcomer.

GenoilGenoil is improving continuously and optimized for Ethereum. It implements easily, and you'll be able to get up and running

with it in. If you are only planning to mine Ethereum, then Ethminer by Genoil are your strong bet.

ClaymoreClaymore is easy to set up and has a ton of Added functionalities, such as enthusiast management, which are not present in

other miners.

That said, it is important to note that there is A very clear difference between

Ethereum mining and mining Ethereum well. The

majority of the applications for mining Ether may carry out numerous calculations simultaneously.

The calculations of these applications may end Up straining the performance of your PC. As a result, prior to making a decision

regarding whether you would like to mine Ether or not, you need to ensure that your computer has adequate processing power

capabilities.

It's also vital to note the Ethereum mining Performance is usually measured in Hashes a second (Hash/s). A Hash per second only

indicates the amount of times that the chip can convert the data that is provided to blocks in one second.

Next up, let's explore the Ethereum mining tools.

Ethereum mining instrumentsHere are a Few of the Ethereum tools that you Need to have before you begin mining:

Ethereum WalletWhen you Start mining, then You'll Need to store Your own Ether in a safe location. It is possible to store your

Ether in two ways, either using a local wallet or an online wallet. A neighborhood wallet has better security since it will always

stay under your control. However, should you use a local wallet, then you have to either install it on a pc that's not your

Ethereum miner or regularly transfer your funds elsewhere.

The main reason for this is that if your Computer crashes, it may be hard to recover any Ether that is stored on it.

Mist / Ethereum WalletBoth the Mist and the Ethereum Wallet are Official developments from the Ethereum Foundation team. While in the center of it, these two systems are straightforward to use and pack a lot of additional features. Since it's integrated with

ShapeShift, it is going to make it possible for you to accept payments from Bitcoin along with other alternative crypto

currencies. If you're considering constructing intelligent contracts, then Mist/Ethereum Wallet is your very best option.

Geth (with Etherwall)Geth is the underlying application code for your own Mist wallet and is the primary service for syncing the

Ethereum Blockchain. Regrettably, it uses commands that are typed on command prompt, making it hard and annoying to use. Geth

(using Etherwall) integrates GUI front-end to it, which makes it easier to use.

My EtherWalletMyEtherWallet is an open source and Client-side Ether wallet which runs on JavaScript. It makes it effortless to

develop secure pockets without using the command line or implementing an Ethereum client on your PC. When you operate

MyEtherWallet on an offline computer, you may easily develop protected paper wallets for your Ether holdings.

Poloniex and KrakenPoloniex and Kraken are online crypto money Trading platforms. You can use the deposit speech at Poloniex and

Kraken to transfer any Ether which you make directly.

Preparing the mining procedureWhen it comes to selecting your Ethereum Mining OS platform, there are 3 chief contenders: Windows

OS, Linux OS, and ethOS. Let us jump in and explore the way the process of preparing Ethereum on these platforms.

Mining Ethereum on Windows OSHere are the steps that you can follow along with mine Ether onto a Windows OS:

· Download the Geth applications application. Geth is an application that acts as an interface between your computer and the

Ethereum network.

· Unzip the Geth file on your own PC.

· Run the program that you've just downloaded. To do the program that you have just downloaded, type in the search button that the

keyword "cmd" and click on it.

· Once the command prompt is started, type in "cd /" to navigate to "C :\>."

· Create a new account by typing "geth account fresh" in the command prompt and press the enter key. You should have a screen like

the one under:

· Type in the password that you'll be using in your Ethereum network

·	Allow the Geth application start communicating with the rest of Ethereum system. To attain this, type in "geth -- rpc" at the

command prompt and press the enter key. You should see a screen similar to the one under:

· Await the Ethereum Blockchain to finish downloading and synchronizing with the Ethereum program.

·	Download and install the Ethminer program. The Ethminer enables your CPU to run hashing algorithms in a fast and safe method.

·	Open another command prompt and browse to "C:\Program Files" by typing the "cd" command in the command prompt.

· While in "C:\Schedule Files," type "cd Ethereum 1.0.1\Release" and press the enter key where 1.0.1 is the version of the

Ethereum miner. Obviously, you will have to confirm this first.

· Now type "ethminer --G" at the command prompt and press the enter key. Wait for the mining procedure to start. The DAG file will

be created over 10 minutes. You should now see a screenshot similar to the one under:

Mining Ethereum on Linux OSHere are the steps Which You Can use to mine Ethereum on Linux OS:

1. Step one: Ensure that you have installed and upgraded your AMD Driver to the latest version. You can now download the

AMDGPU-Pro driver, unzip it and install it.

2. Step two: Install the Ethereum Software. To install the Ethereum customer, you will have to bring the repository. If you're

using a Debian-based Linux distribution, like Ubuntu, type the following commands in the command prompt of your Terminal:

Sudo apt-get set up software-properties-common

sudo add-apt-repository ppa: ethereum/ethereum Sudo apt-get update

If you are using a CentOS/Redhat OS, for example Fedora, type the following commands in the command prompt of your Terminal:

sudo yum install software-properties-common

sudo add-yum-repository ppa: ethereum/ethereum Sudo apt-get update

3. Step three: Now install your favorite Ethereum software by typing the following commands at the command prompt if you're

utilizing a Debian-based Linux distribution OS:

Sudo apt-get install ethereum

Sudo apt-get set up ethminer sudo apt-get Install geth

If you're using a CentOS/Redhat OS, kind the Following commands in the command prompt of your Terminal:

sudo yum install ethereum

Sudo yum install ethminer sudo apt-get install geth

Mining Ethereum on Mac OSHomebrew is by far the easiest way to put in Go-ethereum at Mac OS. In case you haven't installed it on

your own system, install it first. To install Ethereum in your Mac OS, follow the steps outlined below:

· Run the following commands to set up geth:

brew tap ethereum/ethereum

brew install ethereum

· Now You Can set up the develop branch by running the following commands:

brew install ethereum --devel

· Now, run the new geth accounts to create an account on your node.

Ethereum mining profitability calculatorNow that you have set up your Ether and you are Ready for mining, what is next?

It is now time to Begin reaping from your investments in Ethereum. But maybe not that quickly. How do you decide if mining the

Ethers is rewarding or not? That's where the Profitability Calculator comes in. The Ethereum mining gain calculator can help you

to find out if mining Ether is going to be rewarding. The Ethereum Profitability calculator gives an estimate of the expected

crypto currency earnings according to statistical calculations.

These statistical computations are performed using The values entered and does not account for the difficulty and exchange rate changes, the rancid or rejected rates, along with the mining pool's efficiency. If you're mining using the pooling system, the

estimated expected Ethereum earnings can vary depending on the pool's efficiency, stale prices and the fees charged. On the other

hand, if you are mining utilizing the solo approach, the projected expected crypto currency earnings can vary depending upon your

fortune and the stale or rejected rate.

For assistance with some of the calculations Ethereum Miners need to make, there are many sites which supply Ethereum

profitability calculators. In all these calculators, all you need is to enter parameters like how much your equipment cost, hash rate, power consumption, and the present Ethereum price to observe how much time it will have to have a greater ROI.

Ethereum Virtual Machine (EVM)The source code for Bitcoin was implemented in C++. Hence, the restricted scripting for Bitcoin apps

often happens at a more compact layer. The internet impact of granular programming is that it makes the system less desirable for

the development of several new-age web-based systems. However, the high-level languages that are easily obtainable inside the

Ethereum ecosystem create smart contracts available to most developers.

For Example, Geth--that can be executed in Go-- Is available to Ethereum programmers and can be used to encode complicated smart

contracts in a simplified manner using the EVM. In other words, Ethereum is a more expanded form of Bitcoin built not just as a

crypto currency system but with intelligent contracts in mind. That is why any programmer interested in the future evolution of

smart contracts should think about knowing the EVM.

But What's the Ethereum Virtual Machine?

Ethereum is a programmable Blockchain system. As opposed to just providing you with a pair of pre-defined surgeries--such as those

in Bitcoin transactions--Ethereum allows you to create your own apps of any size and sophistication. Specifically, Ethereum serves

as a digital platform for the development of decentralized apps, including, but not limited to crypto currency systems.

Ethereum was invented with intelligent contracts in mind. At the heart of Ethereum is your Ethereum Virtual Machine which is just

abbreviated as EVM. EVM can do the smart contract code of any arbitrary algorithmic length and sophistication. In technical terms,

EVM can be perceived as a whole "Turing machine".

The Turing machine could be viewed as a "world computer" that any developer can create and run programs on using a friendly

programming language. For instance, you may make wise contracts and execute them on EVM using programming languages like Python,

Go and JavaScript. Sounds interesting, isn't it?

Now, just like any Blockchain-based system, Ethereum also utilizes a peer-to-peer network protocol in which the Blockchain

database is supported and updated by multiple autonomous nodes which are linked to the Blockchain network. Each node of this

Blockchain network must run the EVM and execute exactly the same intelligent contract directions.

Now, to Start creating your smart contracts, You'll require a client that connects to the Ethereum network. Specifically, the

client will serve as your window into the distributed Blockchain system and provide you with an opinion of the Blockchain-- where

the EVM has been set up.

You can use Solidity, Serpent or Pyethereum Platforms to execute your smart contracts. Let us jump in and explore these programs.

SolidityUndoubtedly, Ethereum would be incomplete Without a native programming language. That language is none aside from

Solidity. Solidity is your code that allows Ethereum developers to run programs and contracts in a decentralized way. It looks

like the browser-based JavaScript language but is instead used for running Ethereum smart contracts.

In contrast to Object-oriented languages such As Java and JavaScript (that combines factors, data, and works to execute certain

human-operated orders), Solidity is a "contract-oriented" language. Its run-time environment tasks are all automated and the items

are bundled together to eliminate theneed for manual commands.

It is often called the Ethereum's scripting language. However, it's actually a language that is compiled rather than a scripting language. It compiles the wise contract instructions into bytecode so that the EVM can browse them. This is a very important

quality of contracts since they aren't independent programs but rather partially compiled codes which rely on EVMs to operate.

Solidity has also been designed to express Agreements that could encode arguments and relationships that normally exist in real

life. Therefore, it comprises more theories than an Object-Oriented language. The identity, ownership and security frequently form

a core part of the Illustrator training.

Since the programming language evolves and adds More libraries and users, it has the promise to create massive and powerful

constructs that may end up having real-world applications, for example IoT devices.

SerpentThe serpent is among those high-level programming Languages which are used to compose Ethereum smart contracts. The serpent

is designed to be like Python. It's meant to be optimally clean and straightforward by blending many of the efficiency benefits of

low-level language with ease-of-use in its own programming style.

At the same time, it provides unique Domain-specific features for smart contract programming. The current version of this Serpent

compiler, which is available on GitHub, is composed in C++, enabling it to be readily integrated in any client. Along with

providing efficiency while programming, Serpent has the following crucial distinctions when compared to Python:

· Even though Python integers have an infinite dimensions, the Serpent integers go around 2256.

· While Python program allows decimals, Serpent doesn't encourage decimals.

· Python compiler supports lists, dictionaries, and other advanced data structures while Serpent has no list comprehensions.

· Python compiler supports first-class functions while Serpent doesn't.

PyethereumPyethereum is the Python core library for Execution of Ethereum intelligent contracts. It gives the basic classes and

necessary routines for interacting with Ethereum smart contracts. We have made a digital machine that contains all the essential

applications. Specifically, Pyethereum permits you to program, letting you interact with all the Blockchain to run and test your

smart contracts.

Pyethereum is a Python-based Ethereum client system. Nevertheless, it isn't the only one available. Today, you will find Ethereum

implementations In C++ (CPP-ethereum) and Go (go-ethereum or just Geth). At the most Basic level, you may start using Pyethereum,

which is the latest OS (whether Windows, Linux or Macintosh), and the most recent version of Serpent.

Ethereum Supporting Protocols

Besides the main Ethereum Blockchain protocol, Other supporting technologies in the Ethereum development seek to complement the

network, which makes the Ethereum parts run more efficiently.

For Example, a whole new set of protocols is Currently being developed to improve the functionality of their decentralized

software while tools keep on evolving to allow the applications to exploit data from multiple Blockchains. While there may be a couple of protocols that combine the Ethereum concepts on the surface, all are aimed at creating the Ethereum more flexible for

developers and users.

Here are some of the most Frequent encouraging protocols:

WhisperWhisper is a communications protocol and Toolset that enables apps built on the Ethereum protocol stack to speak to

another. It combines all the aspects of the distributed hash table and point-to-point communications system to permit apps on the

Ethereum platform to communicate with one another.

For Example, the whisper protocol may Facilitate the exchange of information by buying and selling supplies to provide seamless

communications akin to that of chatroom-like apps. It's possible to consider Whisper as an Ethereum app for whistleblowers who

possess a trove of data stored and may want to communicate with a journalist, but he/she does not want hisor her identity to be

connected to the information.

SwarmSwarm--or SwarmHash as it is commonly known As--is a peer reviewed file-sharing protocol that's intended to store and

retrieve data efficiently to be used in Ethereum programs and contracts. The simplest analogy to draw with Swarm would be that it

is basically the BitTorrent to get Ethereum.

There is no doubt that to store information directly On the Ethereum Blockchain could be costly. While the contract code is to be

kept on the Blockchain, any mention information that's necessary for contract implementation should not. As an example, if a

simple, intelligent contract were to state deliver an e-card with the pictures, the photographs would take up a vast volume of

space.

Possibly a school Might Want to send out a record With photographs of its graduating class. If such an application runs on

Ethereum, it may demand a contract that is 1 KB but is intended to deliver approximately 1 GB of data. Storing and transacting a 1

KB of code may cost users a couple of bucks, whereas keeping the record itself will always be more expensive.

By keeping the album remotely and obtaining The file using a BitTorrent-like system like Swarm, it might allow Ethereum apps to provide the instructions, together with the documents to be moved, with the Swarm system and not the Ethereum Blockchain.

OraclesTo get Ethereum contracts to execute correctly, They need not only a well-designed series of "if ...then" statements, but

also a way of determining the accuracy of the given inputs to people "if ...then" statements. Suppose it is raining in Singapore

and many reliable sources can verify that it's raining, but how can Ethereum weed through potentially fraudulent sources to spot

the accuracy of the inputsignal?

Ideally, we would have a build that Communicates the outside realities to the smart contracts. Back in Ethereum, the constructs

that describe the external truths are known as oracles. While a number of the jobs are trying to create their own private oracle

systems, there were efforts to develop one system to help confirm inputs to multiple Blockchains.

Even though in the moment, you will find a restricted Number of data sources, creating a unified framework is achievable and can

be cryptographically proven. In fact, it is not hard to figure out a future where smart technologies and the Internet of Things

will allow all kinds of outside information to be integrated into the wise contracts.

MistIf the Ethereum protocol would be the brand new TCP/IP That encourages sharing of value instead of just storing information,

the job needed for the new version of the browser which provides a beneficial frontend technology is unquestionably Mist. Mist provides users with capabilities to explore the apps and offerings that use the Ethereum protocol.

Styled as a distributed app discovery tool, Mist is intended to serve as a pocket for the smart contracts that allows GUI-allowing

users to place their trade fees and handle customized tokens dynamically.

Chapter 6: What is Ethereum mining

Mining is a computationally intensive work that requires a lot of processing power and time. Mining is the act of participating in a given peer distributed cryptocurrency network in consensus. The miner is subsequently rewarded for providing solutions to challenging math problems. It is done by putting the computer's hardware to use with mining applications.

All the information on cryptocurrency transactions must be embedded in data blocks. Each block is linked internally to several other blocks. This creates the blockchain. These blocks must be analyzed as fast as possible to ensure a smooth running of transactions on the platform. However, the issuers of such currencies do not have the processing capabilities to handle this alone. It is where miners come in.

A miner is an investor that devotes time, computer space and energy to sorting through blocks. When the mining process hits the right harsh, they will submit their solutions to the issuer. After verification, the issuer of the currency offers rewards which are portions of the transactions they helped in verifying. They also offer digital coins in exchange for the work of miners. The result of digital mining is called proof of work system. Some currencies depend on this system alone while other use a combination of proof of stake and proof of work.

Mining is a word that originates from the gold analogy of the cryptocurrency sphere. It is not some get rich quick scheme. It requires time and effort to grow especially when you are working alone. The word was adopted because just as precious materials are difficult to see, so are digital currencies. Since mining must take place to increase the volume of precious metals in the market, digital mining must take place to increase the digital currencies in circulation.

The same thing applies to Ethereum. The only way to utilize Ethereum is with the product from mining. However, mining

Ethereum means more than increasing the volume of Ether in circulation. It is also necessary for securing the Ethereum network as it creates, verifies, publishes, and propagates blocks in the blockchain.

Ethereum Mining is the process of mining Ether. Simply put, mining Ether equals securing the network which in turn ensures verified computation.

Ether is an absolute essential, as it serves as fuel for the smooth running of the Ethereum platform. An interesting way to look at Ether is an incentive used to motivate developers to create top notch applications.

Every developer seeking to engage and make use of smart contracts on the Ethereum blockchain needs Ether to proceed. It is popularly called the fuel that runs Ethereum. It is a less expensive way of running transactions on the network when compared to buying Ether. You can also decide to sell your Ether after mining.

Ether supply is not infinite. The overall amount of ether and the network operations was decided at the 2014 presale. No more than 18 million Ether gets issued every year, which is about 25 percent of the first issue. It serves as a system to reduce inflation.

Each block must have the proof of work of the given difficulty if it is too validated in consensus. The algorithm for validation is called Esthash.It has to do with identifying the nonce input to the result in such a way that it will be below a threshold that is determined by the difficulty. If the outputs are uniform in distribution, then the fact that the time required to find a nonce depends on the difficulty is guaranteed. In this case, simply manipulating the difficulty will allow a miner control how much time is required to find a new block.

In Ethereum Mining, the difficulty is adjusted dynamically so that the network produces one block in every 12 seconds on an average. Thanks to the synchronization of the system, it is not possible to rewrite history or maintain a fork except the individual attempting to do so has over half of the mining power in the network.

Ethereum Mining can be done in the comfort of your home. It requires script writing and some knowledge of the command prompt. It is quite easy and exciting once the process is broken down into manageable steps.

Before learning the steps, here are some basics you should keep in mind:

THE BASICS

Mining Ether takes up a lot of electricity. On the plus side, though, if mining practices are carried out efficiently more income is generated through the sale of Ether. Ethereum mining calculators are available for calculating profits. So, there's no need to get worked up since you'll still get a profit at the end.

You can use any personal computer to mine Ethereum, provided the system has a Graphic Card (GPU) with at least 2 GB of RAM. Central Processing Unit (CPU) mining is simply an exercise in frustration. It takes an extended period to complete, and the profits are little thanks to the cost. GPUs are your best bet as they are 200 times faster than CPUs when it comes to mining Ether. AMD cards are more efficient than Nvidia cards as well.

Before following the provided steps, outlined below are some essential information you should keep in mind:

Mining Ether takes up a lot of electricity, and you are right in being concerned about this. On the plus side, though, if mining practices are carried out efficiently more income is generated through the sale of Ether. You will need your Ethereum Mining calculators for the process. Ethereum Mining calculators are available for calculating profits. So, there's no need to get worked up since you'll still get a profit at the end.

It is imperative to have a lot of free space on your computer system's hard drive.

About 30 GB should be sufficient for the blockchain and other software.

THE ETHEREUM MINING PROCEDURE

Step One

You will need to Download Geth. This application will serve as a communication hub, linking you to the Ethereum platform while coordinating your setup (hardware and all) and reporting any new development that requires action on your part.

Step Two

Geth usually comes as a zip file, unzip and transfer the file to the HDD. The C: drive is usually best for this.

Step Three

You need the Command Prompt to execute the installed application. Search in Windows for 'CMD,' if you are unsure about this, then click on it from the search list.

Step Four

'C:\Users\Username>' the username placeholder is the name of your computer and is the usual display format by the command terminal. Locating Geth is the next step; type in 'cd/' into the command prompt terminal; this is an instruction to change directory. 'C:\>' should be highlighted now, which means you are currently in the C: drive.

Step Five

Account creation comes next. To make a call to Geth; type in 'geth account new' followed by the enter key. The command terminal should now display 'C:\>geth account new.'

Step Six

A password will be of you at this stage, and extra care should be taken here. Be sure about your password, write it down if possible and be sure to type it in carefully. Press enter once again after typing in the password and voila! Your new account is created.

Step Seven

Geth needs to link up with the network before anything becomes fully operational. Type in 'geth —rpc' in the terminal and then press enter (you should be used to this by now), this action starts the download of Ethereum's blockchain and synchronizing with the global network. This process is time intensive and is dependent on how large the blockchain is currently and the speed of your internet connection. Make sure you wait until the completion of this process before mining.

Step Eight

To proceed you need a mining software, which aids your GPU in running the hash algorithm required on the platform. Ethminer is a good choice for this heavy lifting.

Step Nine

Install Ethminer or any other mining software you choose for this procedure.

Step Ten

Repeat step 4 in a new command terminal (change directory command). To open a new command terminal, right-click on the previously active terminal icon found in the taskbar and then click on the terminal from the menu.

Step Eleven

In the new terminal window, type 'cd prog' followed by the tab key. 'C:\>cd prog' should now be on display, press the tab key again to display 'C:/> cd "Program Files"' then push the enter key to show 'C:\Program Files>.'

Step Twelve

To go into the Ethereum mining software folder, type 'cd cpp' then the tab and enter keys. Press tab once again and the terminal should now display 'C:\Program Files\cpp-ethereum>.'

Step Thirteen

To start mining with your GPU, key in 'ethminer –G' followed by the enter key.

This will initiate the mining process after building the DAG ((Directed Acyclic Graph), which is a large file stored in the RAM of your GPU for the purpose of making it ASIC (Application Specific Integrated Circuits) resistant. Ensure that there is sufficient space on your hard drive before getting to this point.

Step Fourteen

Conversely, if you are up for it, CPU mining can be done. Just type in 'ETHMINER' then the enter key to start the process. The building of a DAG is still required in this step after which Geth takes over communication with Ethminer.

Chapter 7: What Makes Ethereum Different?

In this chapter, we're going to be discussing many of the neat features which are associated with Ethereum. If you're interested in Ethereum, then the chances are that you've already heard of one or two of these features. These are the things which drove the immense success of this cryptocurrency and are the things which will carry it forward further as time bores on. The future of Ethereum, coincidentally, is something that we'll be discussing in a later chapter.

So what exactly makes Ethereum different from other cryptocurrencies? To understand this properly, we have to look back at the development of it and why it was created in the first place.

Ethereum started to be developed early on in 2014. It was the brainchild of Vitalik Buterin. It was the natural end of a suggestion by Buterin that Bitcoin would fare better were it to have a native scripting language.

The Bitcoin community, in general, didn't quite agree with his point, however. The idea failed to catch steam at large. As a result, Buterin decided that a good solution would be to start working on another platform much like Bitcoin, but this cryptocurrency would utilize a unique scripting language.

This is the core part of Ethereum and what really separates it from other cryptocurrencies: it was the first major cryptocurrency to allow for the integration of scripts into the blockchain. In the first chapter, we talked a bit about the blockchain, what it is, and how it can be used.

The idea of Buterin was to build onto this concept by allowing different scripts to be embedded into the blockchain. These scripts would then allow many things that weren't already automated to be automated with relative ease. This would be a major upset to the cryptocurrency industry at large because it began to encroach upon a whole new frontier that people hadn't considered before; it also allowed for the possibility to displace professional mediums such as lawyers that would usually act in mediation between two parties attempting to carry out a certain agreement.

These scripts, which can be embedded in the blockchain and constantly run, are referred to as **smart contracts**.

Smart Contracts

Smart contracts are one of the most pivotal pieces of technology within Ethereum. There is no doubt that without the institution of smart contracts, Ethereum as a whole would have had a much harder time taking off or, more likely given that the ability to develop scripts and integrate them into the blockchain was a central part of the development of Ethereum, would have even existed at all.

Smart contracts are very simple in nature, but they are fantastic because their execution can really be infinitely complex. Smart contracts are simply scripts with sets of terms in them. These terms can then be constantly evaluated, or they can also be set to trigger only when a certain condition is met. This allows for a great many possibilities.

For example, let's say that there was an Ethereum-based asset trading marketplace. In this particular example, we'll use stocks. Let's say that you are trying to buy and sell stocks using Ethereum as a currency. Ordinary currencies fall flat because they don't allow for automation.

As an example, let's say that you had 600 shares of stock A, and each share was valued at $3. Using standard currency, you are at the whim of your own discretion. What I mean by this is that the currency itself cannot automate the process of buying and selling. Sure, there are many marketplaces that offer in-depth user tools which allow you to automate the process, but those are a feature of the **marketplace**, not the **currency itself**. This offers a great many difficulties. For example, if the marketplace were to experience difficulties, then the automation may not execute correctly, and you may miss your opportunity to sell at a certain price point because of a failure on the technological end. This would leave you at a market disadvantage because you'd be the only person for whom the technology failed (alongside others who used the marketplace and had the same technological error, but I digress).

However, through the usage of smart contracts, you can very easily set up the automatic selling of your shares of stock A when the shares reach a certain point - let's say $3.50. This can be integrated into the blockchain which means that for as long as the blockchain persists, your script is guaranteed to execute. It also adds another layer of security because there is no abstraction there; you know exactly what the script to be executed is.

This idea of smart contracts was and continues to be the central thing which divides Ethereum from the competition, and though some have attempted to implement such a technology since then, none have met the same degree of mainstream success that Ethereum has.

Ethereum Virtual Machine

All smart contracts are run within what is referred to as the Ethereum Virtual Machine, or EVM. If you have no programming background, then virtual machines are essentially things with which programs can run which abstracts them from any other processes on the computer. The popular programming language Java is run within a virtual machine.

All Ethereum scripts run with Ethereum's virtual machine. This means that no scripts running from the Ethereum blockchain may end up harming your computer because it is completely and totally isolated from all other factors on your computer. It also means, per necessity, that many more platforms can support Ethereum scripting because all that must happen is that a version of the Ethereum Virtual Machine must be ported to the platform. This also means that many different platforms can easily run an Ethereum node with ease. (We've already talked about what nodes are, in chapter one.)

Programming in Ethereum

So we've talked about the scripts and the nature of Ethereum, but we haven't really covered programming in it, all that we've covered is that it's possible. We're not going to go into incredible depth in this section, but we are going to talk about it a little bit. There are more specialized books if you are interested specifically about learning programming for Ethereum.

All that must happen for a language to be compatible with Ethereum is that it must be compatible with the Ethereum virtual machine. There are three different languages which have received large-scale implementation.

The largest is Solidity. Solidity is the premier language for Ethereum scripting. It is based on C, C++, and JavaScript, and therefore bears many similarities to C-style languages. It's not terribly difficult to write in Solidity if you have familiar with other languages because many of the conventions will be the same.

Another is Serpent. Serpent is increasing in popularity due to its immense ease of use. It is ultimately based on Python and is very similar to it syntactically, which means that people who are new to programming can still pick up Serpent programming with relative ease. It doesn't take a whole lot of background to learn Ethereum scripting using Serpent.

Additionally, there is LLL. LLL is the third major language in usage for Ethereum scripting. It is a low-level language, which means that many factors aren't automatically performed where they are in the other languages, and it's ultimately similar to Lisp. If you've got familiar with the Lisp language, then LLL will be very comfortable for you. However, if you lack familiarity with Lisp, then much like Lisp itself, you will not feel comfortable working with lisp.

Programming Ethereum scripts are really easy, for the most part. The logic doesn't differ from other programming languages. The only thing which may throw you off a bit is that you're working within an entirely new paradigm which you aren't used to at all. However, once you get used to the learning curve of Ethereum, you'll have a pretty easy time with it.

What makes Ethereum better?

At this point, you may be wondering, what makes Ethereum the future? What makes it better to other cryptocurrencies? Well, the fact that it has so many uses. We'll be going over this in the next chapter a bit more in-depth, but Ethereum has pretty much an infinite ceiling for what it can be used for. More than that, the development team for Ethereum has shown time and time again that they aren't afraid to innovate.

Ethereum was released only in 2015, yet it still has a firm grip on second place in terms of market share and only seems to be catching up to Bitcoin every day. All signs indicate that if there were any cryptocurrency poised to catch on in the mainstream, it would be Ethereum.

Chapter 8: Investing in Ethereum

Now we will speak about my favorite part, the investing. As with any investment, what you are doing is essentially buying something to maintain or increase in value over time and that the stability in value or growth will be greater than stuffing your money in your mattress. Just like Bitcoin, you will exchange your currency for Ethereum which will be produced in your secure wallet. Ethereum is new, and an evolved form of Bitcoin and thus has differences that the new user will not be accustomed to if they have used Bitcoin in the past.

When someone bought their first Ether on January 26th, 2016, they had to pay a just a little over 0.006 BTC for one Ethereum. After just two weeks, this price appreciated in value to the tune of 0.017 BTC as of the second week in February. That's a 300% return in just two a few weeks' time making this Ethereum currency look like a very lucrative part of any portfolio.

That initial Ether bubble burst shortly afterward but has continued to grow, as of September 17th, 2016 the price for one Ethereum is 0.0208 BTC, and it remains the second largest CC as far as market capitalization, with Bitcoin in a strong lead. This is an impressive result for an 8-month old CC and with the benefits outlined above; there is clearly a larger market to be captured, and a larger growth to be had in the future.

We need to discuss how Ethereum actually works. That is, how can we possibly make money off of this? What causes the price in Ethereum to rise or fall, and how can we understand this in a way to properly make sound investment choices?

As we stated above, the value of Ethereum is through its contracts. A simple example contract would be one that requests a specific user to sign into a website a certain amount of times. Once that has been completed, the conditions of the contract have been met, the payload will be released, and the transaction will execute with payment into the contract "signer's" account. The payment is sent out autonomously, and either party requires no other action.

Thus you can see the benefits of this, as opposed to say using PayPal or any other payment system where you would likely have to show proof of visiting the site the appropriate number of times; the payer will then have to send funds manually. These funds may not even be immediately available due to "settling times" which are finished within a few seconds with Ethereum – but can take days with traditional banks.

In August 2015 the original distribution of the Ether coin commenced. It was the product of a crowd-funding campaign that lasted just over a month and resulted in a staggering $60 million dollars being reached – the largest ever crowd funded project on Earth!

Some of the major ideas that are likely going to be behind the likely rise of Ether prices have to do with overhauling certain industries, making them more efficient as well as opening up more streamlined peer-to-peer options. An excellent example is the stock market.

Currently, in trading within the stock market, there is always at least one middle-man, the market-maker(s) (MM). While MMs provide the market with liquidity, additional middlemen are not required and more often than not, result in excess fees and deficiencies in making the buying and selling of stock quick and painless. Ethereum would be able to replace your broker in this situation, thus removing the sometimes massive fees associated with buying and selling stock. This is only possible through the use of the smart contracts. The MM, with Ethereum, could produce contracts that offer any user with Ether to purchase stock directly from the MM – this is currently an impossibility for most Americans, who are **required** to go through a broker.

The above is an example of how you would be investing (buying stock) **with** Ether, but it is a reason to invest **in** Ether as well. Any benefits that Ether can produce in comparison to regular currency or BTC will cause demand for this currency to increase, and its value will necessarily increase as well. You should also be aware of all the major projects that are being looked at within the Ethereum community. If any of these projects are massive breakthroughs and would make things more efficient for people or create a new demand for Ethereum, this would be an impetus for you to be investing in Ethereum. We can look at some of these projects that are going on within the Ethereum community now.

Augur is a Dapp and can be likened to an online gambling platform. This Dapp is going to change the way people make bets on certain outcomes in a dramatic fashion based on its design and broadness in capabilities. Some liken Augur will do to the neighborhood bookie what the electric freezer technology did for the old ice delivery man.

The purpose of Augur is simple but powerful. It will allow any Ether holder to wager money on **any** future event of their choosing. If the event doesn't yet exist, anyone can make an event for people to place their bets on. Software built on "wisdom of the crowds" will set the probabilities, collect the Ether bets, and finally, reward the winnings. Because of the ease in which this can be set up, the price for actually running the show is exceptionally smaller than that compared in the average bookie. Only one percent is going to go to the Ether bookie, which is ten times lower than the average today if you go to Vegas.

However, the design of Augur isn't meant to be a Vegas casino. It has no interest in roulette, blackjack, or poker. It's not so much built on "random" probabilities but rather relies on the idea of "wisdom of the crowds." This is exceptional for binary events such as sports betting, political race outcomes, market forecasts, even weather.

What's truly novel about Augur is that it is not going to be controlled by any single person or entity, and it will not operate on any single computer network. All of the currency within the Augur system will be in Bitcoin or Ether (or other types of cryptocurrency), so no central credit cards or major banks are going to be involved. That makes the legality of the system completely up in the air. If the system becomes successful, regulators and bookies will organize to sue and undermine the system. The problem is, there isn't anyone to point the finger at, as the creators, as well as the users, are all decentralized, and in almost all cases, anonymous. Thus it's success will only bring more attention and success to it.

This is a new legal territory, and we will continue to see new legal territories develop with these decentralized networks. The question is if Augur becomes exceptionally successful for betting on the horse race or betting on the Dow Jones a week from now, can anything be done to stop it? The answer is, at the moment, no. If action were to be taken on Ethereum, it would shift the legal power of any government involved, and that is going to take a lot of time, a lot of energy, and a **lot** of resistance. So for at least the next few years, it's likely that Augur will remain up and running, if not indefinitely.

Augur may be destined to become the web's answer to gambling prohibition—it will do to the betting man what the silk road did to the drug user—but you'd never know it from talking to the developers of the system. They aren't a shady bunch at all, working in San Francisco, attending conferences, and have well-built legal representation in the real world while talking openly about what they're producing to any media outlet interested. There's even an infographic movie funded by the group. As you may begin to see, the confidence in the decentralization of these networks, allows the developers to do what they will, and the **use** of such tools in the crypto-world makes it difficult for regulators to write up a document that could put a stop to it.

Around $300,000 of the total $600,000 that was raised by Augur's funding team comes from a man named Joe Costello. Costello is a successful tech entrepreneur, known to be one of Steve Jobs' top picks for the new CEO position of Apple itself. Following the smart money isn't always a dumb idea.

Gambling or casino are terms never used by Joey Krug, a young Pomona college dropout, but also Augur's lead developer. He and the small team of just five employees use the term "prediction market."

Due to Augur being based on the wisdom of the crowd, as bets are moved in and out of each outcome, the odds will appropriately adjust. This is very interesting because not only does it perfectly price the outcome (based on the weight behind each choice), but it also generates a crowd-sourced statistic. There are already groups and programs that use this crowd-sourced mind, such as UNU. UNU used crowd-sourced (what they call swarm intelligence) ideas to predict the superfecta in the Kentucky Derby correctly. This is correctly predicting the first four horses, **in the correct sequence**, to cross the finish line. Another example, InTrade, which was the most frequented prediction-market up until federal regulators shut it down for U.S. customers in 2012. The "swarm-intelligence" seen there beat the pollsters and pundits by correctly predicting the outcome of the 2008 presidential elections in nearly all (48 out of 50) states. Not only would you be investing in possibly the largest prediction market to date, but also the largest crowd-sourced statistics as well.

What Augur ultimately hopes to accomplish with their platform is to make it possible to do a simple Google search and find the probability of any future event, based on all the participants that took place in the poll. Ideally, this would upgrade our world by creating the availability of more informed policy decisions.

But, what will likely be Augur's most profitable aspect is going to be the cost reducing and convenience aspects in which gamblers, betters, and speculative investors place their bets. Augur actually has the potential to make the world safer by taking away market share in the gambling industry from criminals.

Weifund is another example of something taken from the regular internet and brought to CCs and additionally decentralized. You can think of Weifund exactly like Kickstarter, but it supports any type of CC. There will be ways for Weifund to produce token rewards for redemption in some way or another as well. One of the options through Weifund is shares within a project just like investing. You can then trade these shares on the EtherEx exchange.

Colony has a concept that is centered around the idea of a True DAO. A true DAO is a step beyond a decentralized fund, and into a decentralized company or non-profit. Colony is designed and meant to support a variety of Internet Organizations. Colony will allow its contributors to be reimbursed either in CC or with partial ownership of the DAO itself, which will be proportionate to the value of the contribution.

The Colony project has the potential to bring in the biggest organizations, like banks and financial institutions which require hundreds of thousands, if not millions of small transactions every day that go on in the background. These are meant to make sure everything is running correctly and to verify where money is and where it isn't. Colony works to minimize the overhead expenses of this operation by minimizing the transaction costs but also help remove the actual physical requirements that are needed to house the employees who deal with this back-end banking and financing. Additionally, due to its decentralized nature, the concept of "employee" may even diminish, as contracts for certain tasks can be sent out to be done by literally anyone.

Let's take a look at the total money in Ethereum to understand our upside and downside. The bottom line is that if you purchase Ether, and the market cap increases, the return on your investment will increase. The current market cap of Ether is about $1 billion, while Bitcoin is around $10 billion. The simplest scenario is if you were to buy one Ether token today and if the market cap of Ether went up to $10 billion (that of BTC) you would have made a 10X return on your investment.

The market cap of $1 billion for Ether shows that there is a strong interest in it and that the demand is high. It also indicates approximately, the total amount of money inside of Ethereum (but not quite). With a $1 billion market cap, now large projects, such as banking and financial services can become possible, and the confidence in the system is likely to attract larger players. If or when large banks begin to transfer over their requirements for operations into a cryptocurrency market, the value of that market is going to increase exponentially. When a financial entity moves assets into CC, you can have a great deal of confidence that we will start to see a transition from the fiat currencies to cryptocurrencies.

If the price of Ethereum gets too high, the transaction costs will become too great, and dilution (distribution of Ether) will be required to keep the system running smoothly. This would be a negative outcome for any investment into Ethereum. Financial entities would be wary of this and may want to see stability in the market before they begin to get their toes wet. Ironically, if a financial institution were to broadcast that it was moving a significant portion of its operations into the Ethereum platform, that may actually trigger such a high demand for Ethereum that it requires dilution! The volatility that ensues such an event would be large and would cause any investor to feel some heat from fluctuating prices.

With that said, if you were to be a long-term investor in Ethereum, you have to focus on the long-term timeline, with short-term volatility being a necessary even to find traction within the market, and for value to solidify. Thus, if you believe that the demand for Ethereum will be higher in one, two, five, or ten years (your duration in which you plan to hold), then you need not worry about the fluctuations that occur in the interim.

Let's take a look at the actual steps to investing in Ethereum now that we have some of the theory and the future outlined for us. The first step is creating a wallet. The wallet is going to be your home base, where all payments received are sent to, and all payments outgoing are sent out. The easiest wallet for Ether is likely going to be "Ethereum Wallet," who would have guessed? It's a desktop application and can transfer whatever altcoins you receive into your wallet as Ether, which is very convenient.

Next, you'll need "Geth" – this is a command line interface that is essential for programmers that want to build contracts. If you're not good at programming, you may skip this step, but this is where the most income-generating would come from. Etherwall, a program that will build off of Geth is a way of using Geth to interact with the network with an easy to use interface. This is very beneficial for ease of use for sending, receiving, and creating contracts. Additionally, MyEtherWallet is a go-to for storing multiple wallets and runs on Java. You can use MyEtherWallet offline to make "paper" wallets.

Now filling your wallet is done by two ways initially: mining or buying it straight out. Buying Ether has many options, but you'll likely want to use Shapeshift.io. This is a convenient, easy to use site that requires no actual registration to use. The biggest benefit here is that you can switch between the 33 most abundant CCs currently being used. Create your account with Shapeshift and click Bitcoin for deposit box and Ether for receive box. Use your public address for your wallet depending on the method you used above. You'll likely want to make an initial deposit of Bitcoin into the account, but from here on out, it might be wise to start mining for Ether in the Ethereum system.

There are two major ways of doing this. One is to simply use your own devices to mine Ether, or you can rent cloud-processing power to do it for you. You'll likely want to do the math to make sure this is worth it at the market's current prices, but it's usually around $45 per mh/s.

Genesis Mining is a great choice to start mining Ethereum. It was one of the most socially active Cloud Mining companies in the cryptocurrency community and has a reputation in the field. It has been a major contributor to a lot to the Ethereum promotion process and has actively participated in the bringing the discussion of scaling cryptocurrencies to market leaders. By being open and confident about the success of cryptocurrencies and mining, Genesis Mining has gained an unparalleled degree of trust in the Cloud Mining industry.

To check your balance and send Ether, you'll need to go to your wallet and "View Wallet Details." You're required to submit your private key to access the data within your wallet. It's more or less a password, and should only be known by you. This will allow you to see your public key, balance, and other important information in your account.

Sending Ether is as easy as looking up your wallet details. Just find the button "send transaction" to initiate this. Make sure your wallet has enough funds for any outgoing transaction, of course.

Once you've funded your account, you have officially invested in Ether. You can now speculate on price changes, or you can put the Ether to use by making smart contracts, the heart, and soul of Ethereum.

Chapter 9: How Cryptocurrencies Work

Cryptocurrencies have a simple defining characteristic that revolves around immutability (or 'unchangeability') and security. That is pretty much the basis of cryptocurrency.

Currently, centralized organizations have been putting in tons of cash to improve and maintain their security but even so, they still stand largely defenseless against inside forces just as much as the outside forces. It only takes an individual with the right keys to manipulate things for their own gain. For instance:

A bank typically records its transactions in a ledger to keep track of everything. However, a crafty crook can break into the bank and change some details of the ledger so that when he withdraws more money (than he really has in his account) the following day, the bank will not detect any anomaly in his ledger.

Take this other example:

There is a bank whose branches don't trust each other. They decide to create a system where each branch keeps a log of all their transactions and sends a representative to all the other branches to write down all their respective transactions as well. In case of a dispute, the branches now have the option of comparing their ledgers and determining which one is correct by majority.

To minimize the risk even further (to curb against the risk of having its ledgers changed), the bank decides to have all the transaction values at the end of each page of the ledger, and put them all into a complex equation and then record what they receive for safekeeping in a different document. When the crooks come, they have to change the entries and ensure the alterations they create are still giving the same answer when added into this equation. The thieves take a long time to do that, weeks actually, and the bank knows long before then that something is wrong; it then takes out one of the copies of the ledger and handles things accordingly.

Moreover, the bank decides to do something else to stop the bank itself from scheming against their clients; it offers each branch a reward each time they are at a consensus with a majority of the other branches. They fine each individual branch if they find themselves among the minority. The branches find it better to play by the rules instead of breaking them.

Enter Blockchain

The analogy above relates to cryptocurrencies through a technology called **blockchain** and the variances between the methods of consensus.

Likewise, in cryptocurrencies, we have 'nodes' that represent the competing branches in the example above. Each time they fill out a single page of their ledgers, they have to cross-reference it with the ledgers of everyone else. These ledgers are the 'blockchains' and the pages are the 'blocks'. This complex equation used to solve the value of the transactions on the page is called hashing. Note that it is a lot difficult to solve in reverse as a crook would attempt to do (well technically, hashing is the process of taking an input string, which refers to the transactions, and running it through an algorithm to give out a fixed length output).

Being a "node" is beneficial as every transaction conducted over the system pays a particular fee to the nodes for both processing it and making sure it is secure. If you can recall, I talked about some fines in the bank analogy above; well, the fines depend on the consensus process, which occurs in several types.

Types of Consensus

Let's take a look at two of the most popular types of consensus.

PBFT (Practical Byzantine Fault Tolerance Algorithm)

The creation of PBFT sought to provide a solution to a problem that I could only best present in a famous analogy known as the **Byzantine Generals problem**.

We have a city surrounded by an enemy army known as Byzantine army, an army that has divisions, with each division commanded by a general. The generals can only talk to each other through a messenger. They begin by observing the city and soon, they have to decide on a consensual plan of action.

We have to appreciate that some of the generals might be traitors who are trying to keep the devoted generals from coming to an agreement. There is need for a good majority to attack the city and decide when to do so as well. The generals need an algorithm to act as a guarantee that:

The devoted generals agree on one plan of action

A few traitors don't make the devoted generals take up a poor plan

While the devoted generals will obey the algorithm, the traitors may conversely do what they wish. Regardless, the devoted generals have to agree and come up with a good plan.

Just to clarify the analogy above, the generals are the parties (nodes) who participate in the distributed network that is running the blockchain. The messengers sent back and forth are the means of communication across the network the blockchain is running on. The shared goal of the devoted generals is deciding whether to accept some information submitted to the blockchain as valid or not. If valid, the information therefore means it is a great opportunity to resolve to attack. The devoted generals are essentially faithful participants in the blockchain interested in ensuring the blockchain has integrity, which means only ensuring the acceptance of factual information. The traitorous generals are any party looking to falsify blockchain info. They could have a myriad of potential motives; perhaps a person looking to spend a cryptocoin he does not own or some individual seeking to dodge contractual or votive obligations as indicated in a clever contract he has signed and submitted.

Many different potential solutions are available for the above problem and PBFT has been a good potential solution to create consensus in blockchains. Simply put, PBFT ensures that all 'generals' uphold an internal state—this is ongoing specific status or information.

Upon receiving a message, a 'general' uses it together with his internal state to run an operation or computation. This operation informs the 'general' in question what to make of the particular message. When he makes his own decision about the message, the general shares this decision with the rest of the generals in that system. Ultimately, a consensus decision is usually determined based on all the decisions submitted by all the generals.

Proof of work (PoW)

PoW, on the other hand, does not need all the nodes (parties) to give their individual decisions to have a consensus. Instead, PoW is a system that uses a 'hash function' in order to have some conditions under which one participant is allowed to state his conclusions about the information submitted. All the other participants of the system can then independently verify these conclusions. The parameters of hashing or the hash function prevents any conclusions that may be false.

A node has to figure out a complex arithmetic equation to finish a block (or a page). The purpose of this equation is making sure the nodes work hard enough and in return, receive some currency as reward for figuring it out, something called block reward, along with the transaction fees. This, as you may have guessed already, is what we know as cryptocurrency **mining**.

The PoW nodes can decide to do this by becoming miners (we'll discuss mining details in a later chapter). When a miner gets an answer that is different from the other miners, the system automatically rejects the answer. Thus, the miners have an incentive to be truthful so that they don't waste their time and money for nothing. This means that the only effective way to cheat the system is to control 51 percent of the ledgers (or computing power). It is also very hard to change previous transactions and practically impossible to alter transactions after a few blocks. In any case, if you are a miner with 51 percent of the power, you could be able to potentially stop transactions from confirming and reverse the transactions you make. However, does it make any economic sense to do so considering the total capital investment needed? No.

As you will note in the next chapter, the higher the computing power a node possesses, the more likely the node is to figure out the answer to the equation first before other nodes and thus secure the block reward.

The difficultness of solving the math equation is what those in this field call "block difficulty." If blocks are taking too much time to solve, the difficulty reduces automatically (and vice versa).

Other common/popular consensus methods include the following:

Proof of Stake (PoS)

Proof of Importance (PoI)

Delegated Proof of Stake (dPoS)

In cryptocurrencies, transactions are usually sent from peer to peer from one cryptocurrency wallet to another. For this to take place, the system matches up public codes that relate back to private passwords held by users. These are referred to as cryptographic keys. As you are already aware, transactions between peers are usually recorded on a public ledger of various transactions referred to as a blockchain. Every single user of a particular cryptocurrency can have access to the public ledger containing all the transactions if they download and install a full node wallet (this is as opposed to holding their cryptocurrency on third party wallets like Coinmama and Coinbase). The transaction amounts in this case are usually made public but the person who sent the amount is encrypted. That's not all; every transaction usually leads back to a digital cryptocurrency wallet. The person with the password (known as key) to the wallet is the person said to own the amount of cryptocurrency that is denoted on the ledger. This essentially means that when someone receives or sends cryptocurrency (transfer from one wallet to another) with the right combination of public and private keys, the transaction is often queued up (sequentially) until it is actually added to the public ledger. This is why the technology and the ledger is referred to as block and chain since it is a chain of blocks of various transactions. How does this work?

Well, when a cryptocurrency transaction is initiated, this transaction is sent to every user who has a full node wallet. When that happens, certain types of users, referred to as miners) then proceed to start solving a cryptographic puzzle with the use of specialized software until they find a solution. After finding the solution, this is then added to the block of transactions to the ledger. In this case, the first person to solve the puzzle gets a few newly mined coins as rewards.

With the cryptocurrency craze catching up with the world, different cryptocurrencies exist today. Let's discuss some of the popular cryptocurrencies.

Chapter 10: What is Ethereum Mining?

Before learning how to mine Ethereum tokens (ETH), you should first understand the importance of mining. In a blockchain, before any new block or record is added, the transaction first needs to be verified and confirmed before it can be added to the blockchain. This process of verification is important as you can no longer remove or alter a block once it is added to the chain without affecting all the other blocks. Hence, there is always a demand for miners. Also, take note that you are not mining for Ethereum literally. Remember that Ethereum is the platform; what you are mining for is Ethereum's token known as **ether (ETH).** Let us examine them one by one:

☐ CPU mining

This is also referred to as computer mining. This is where you mine ether simply by using your computer's CPU. This is easy to do and is a recommended mining method for beginners who simply want to experience what it means to actually mine cryptocurrencies like Ethereum. There are several ways of doing this. A simple way is to download the GUI miner known as MinerGate. Simply go to www.minergate.com and follow the instructions for creating an account. Do not worry; it is as simple as creating a new email or Facebook account. You can complete the whole process within a few minutes. You can then use the software to mine ether.

The problem with CPU mining is that you will most probably end up with more electricity expense than the amount of ether that you will earn. This is because CPU mining alone does not have enough hash power to mine a decent amount of cryptocurrency. Another issue that you may encounter is overheating. When you mine using your computer, you should follow a schedule; otherwise, your CPU might overheat and could even be broken.

If you are serious about making a profit by mining cryptocurrency like ether, then you should use a hardware. CPU mining is good only if you starting out and just want to experience mining cryptocurrency, but you cannot expect to make any decent profit from it. This leads us to the next method of mining known as **hardware mining**.

☐ Hardware mining

Since one's computer alone does not generate enough mining power to earn a decent amount of cryptocurrency, you need to use a hardware to allow you to increase your mining power. Take note that when you use a hardware, you still have to use your computer, so you should still be careful with any overheating issues. A good way to prevent this is by following a schedule. Be sure to give your CPU and mining hardware enough time to cool down.

When you do hardware mining for Ethereum, you will have to use a specialized hardware called as **Graphics Processing Unit** (GPU). Before, the developers of Ethereum thought that ether can effectively be mined using ordinary CPUs; but then they later discovered that using GPUs is way better and more profitable. There are many GPUs in the market. When choosing which GPU to use, you need to strike a balance between the high hash rate power (mining power) and electric consumption. You can find many GPUs on sale on eBay and Amazon. Before you purchase any mining hardware, be sure to do your research and check the latest reviews. There are three main drawbacks of using a mining hardware: First, a high-quality mining hardware can be costly. Second, you have to worry about your electric expenses since you will be mining for hours on a regular basis; and third, you will still have to deal with overheating issues.

☐ Cloud mining

Cloud mining is one of the most popular ways to mine ether and other cryptocurrencies these days. With cloud mining, you would not have to worry about buying any hardware or downloading a software. In fact, you do not even have to bother about any overheating issue as you do not even have to use your computer to mine. Instead, a mining company will do all the work for you. All that you need to do is to wait for the mining company to send you ether, which is usually on a weekly basis or as soon as you reach the minimum payout threshold.

Okay, that sounds great. Now, what is the catch? Of course, no mining company would send you cryptocurrency just for nothing. The catch is that you will have to make an investment. Normally, you may see something like this: Invest or pay 1 ether and receive 0.023 ether every week. Although this may seem like an ideal deal, especially in the long run, it is not always that desirable. The problem here is that what the cloud mining company reveals to you is only the **expected** return and not the **actual** return that you will get. Therefore, in our given example, it is possible that you may only receive 0.01 ether in a week, or even lower. It is also common for cloud mining companies to impose an expiration of the contract, while others may allow a lifetime contract. As a rule of thumb, before you invest in any cloud mining company, make sure that you read the latest reviews and read the terms and conditions of the contract carefully. If there is any part in the contract that is not clear to you, do not hesitate to contact the customer support team.

Chapter 11: Smart Contracts

Being a decentralized system existing between the permitted parties, one of the blockchain's most impressive features is that there is no requirement to pay middlemen or intermediaries, which saves you time and prevents conflicts. Blockchains may not be perfect because of internal issues but they are significantly cheaper, faster, and offer more security than traditional transaction protocols. This is the main reason why governments and banks are adopting blockchains at a fast rate.

A cryptographer and legal scholar came up with the idea in 1994 that this decentralized ledger will benefit smart contracts, which are also known as blockchain contracts, self-executing contracts, or digital contracts. His name is Nick Szabo and in the system he devised, computer codes can be extracted from contracts, which can then be replicated and stored inside the system. These contracts are monitored by the system of computers that are running the blockchain. Ledger feedbacks like receiving service or product and money transfer can then be logged.

Smart Contracts Explained

Smart contracts aid the involved parties in exchanging money, shares, property, or anything that has value, through a conflict-free and transparent process while keeping away from a middleman's service.

Think of the technology behind smart contracts as a vending machine. When you need a document, the common process is going to the notary or a lawyer, paying the appropriate fees, and waiting until the document is finished. Using smart contracts, you just put in a Bitcoin or an ether inside the vending machine, which is the ledger, and the document you require is dropped into your account. Furthermore, smart contracts don't stop at defining the rules and the accompanying penalties surrounding the agreement which is what a regular contract does. Obligations are enforced automatically inside smart contracts.

As Buterin explains it, in smart contracts, a currency or asset is converted into code. A program then runs this code which internally validates a given condition which automatically determines if the asset goes to a receiving party or back to the one who sent it, or it can also be a combination of the two. In the background, the document is replicated and stored in the decentralized ledger, which imposes immutability and security.

Let's suppose that you are renting an apartment from a landlord. This can be done by transacting with cryptocurrency and using the blockchain. You will get a receipt which is stipulated in the virtual contract, and the landlord gives you the digital key which you will receive on the date specified. If the landlord doesn't give you the key on time, the blockchain initiates a refund and releases it to you. If the landlord sends the key prior to rental date, the system holds it until the agreed date arrives. That is when the landlord gets the fees and you get the keys.

Simply put, if the landlord gives you the keys, he's sure that he will be paid and if you pay the amount in cryptocurrency, you get the key. The document then gets canceled, and there is no way of altering the code without the parties involved being alerted.

Smart contracts can be used in a variety of situations like insurance premiums, financial derivatives, property law, breach contracts, financial services, credit enforcement, crowdfunding agreements, and legal processes. Here are a few examples.

The Benefits of Smart Contracts

- It gives you autonomy. Since you make the agreement, you don't need a lawyer, broker, or other middlemen for confirmation. Incidentally, third-party manipulation is effectively avoided since the execution process is monitored by the network, instead of one or more parties, that can be biased or prone to error.

- You can fully trust them. The documents are stored, encrypted, and replicated inside the shared ledger. No party involved in the agreement can simply say that it got lost.

- Your document is backed up. A bank can possibly lose a savings account or it can get stolen. Inside the blockchain, all the people involved have your back. The documents are replicated multiple times for backup.

- It is very safe. Websites are encrypted using advanced cryptography methods, keeping your documents safe. They are protected from hackers. It would take someone with abnormally superior hacking capabilities to break in and crack the program.

- It is really fast. Processing documents manually will require a lot of your time and effort due to the paperwork involved. Smart contracts do it digitally by automating tasks using software code, which makes going through business processes much faster.

- It can help you save money. Because you don't need to pay an intermediary, you save money in the process. Using the traditional process, for example, you need to pay notary fees for your transactions.

- It is very accurate. Filling out manual forms is a process that is very prone to error. Smart contracts are cheaper, faster, and error-free.

But Smart Contracts Are Not Perfect

Smart contracts do have some flaws, though. What if there are bugs in the code? Or how are these contracts being regulated and taxed by the government?

Let's go back to the apartment rental analogy. You pay the appropriate fees using the cryptocurrency but before the rental date is reached, the building is deemed condemned. With a traditional contract, the case can be brought to court and a refund can still be awarded to you by the landlord. The blockchain is not concerned with that. It will execute the contract as is.

There are lots of challenges being encountered and cryptocurrency experts are fixing them as they come. Unfortunately, these problems serve as demotivation for potential adopters of cryptocurrencies.

The Future of Smart Contracts

The future lies partially in fixing the issues mentioned above. There are specialized lawyers trying to iron out the kinks in the system.

Changes in some industries, like law, can be triggered by smart contracts in the future. Instead of traditional contracts, lawyers will be producing templates for standardized smart contracts. Industries like accountants, credit companies, and merchant acquirers might use smart contracts in handling tasks like risk assessment and real-time auditing. In the end, it can be a combination of both digital and physical copies, wherein the blockchain verifies the contracts while the physical copies substantiate them.

Chapter 12: Recent Hacking in Ethereum

CoinDash Hacking

On July 17th, 2017, hackers hijacked CoinDash, an Israeli startup trading platform when it was undergoing its initial coin offering. It was the first ICO breach. The startup was planning to raise its capital by selling Ethereum cryptocurrency. When the ICO was starting, a hacker logged into the site and changed the sending address to a fake one, and it was flagged by the company later on. Millions of dollars were diverted to the hacker's address.

As much as the ICO raised $6.4 million, the hacker managed to steal Ethereum worth $7 million before the company pulled the plug during the ICO. This situation made investors less enthusiastic with ICOs. In ICOs, which is similar to stock market IPOs, the main differences include:

Investors get cryptocurrencies and not equity

Offerings are not regulated.

After the hacking was done, a few hours later, Parity released a report saying that the update was done on the code and the wallet was secure.

$32 million lost in Ethereum hacking

Two days after the CoinDash hacking, three wallets from three companies were hacked, and Ethereum that was worth close to $32 million was stolen. This became the third hacking after the CoinDash hack and hacking that occurred in a South Korean exchange called Bithumb, where Ether that was worth more than $1 million was stolen from accounts of users.

 In the same week that the $32 million was stolen, a security alert was issued by a Parity, a smart contract development company, warning people about the Ethereum wallet software vulnerability. Attackers explored the vulnerability to achieve this heist.

White Hackers coming to the rescue

Over $75 million worth of Ethereum was drained to secure locations by white hackers as a way of exploiting the vulnerable wallets to protect the coins from being stolen by black hat hackers. The coins were held by the White Hat Group until the threat was eliminated.

Chapter 13: Analyzing the Market

Most newbies in the crypto world are getting into the market because they have heard many rumors about how much money there is to be made in this new technology. They've heard about the incredible profits of Bitcoin and those that follow along with it. Chances are if you're reading this book, you're doing so because of some rumor you've heard.

In most cases, they didn't have a definite plan. They just wanted to get in the market, buy some Ether, and wait for the price to rise. That is definitely one way to go, but there is an awful lot more than that to think about.

While there are quite a few websites that will do the analysis for you, it is extremely important that you at the very least learn to do some form of analysis yourself. There are two primary forms of analysis you can do before you decide to invest. A fundamental analysis involves getting a closer look at how the coin makes use of the Blockchain technology it is using.

Basically, we want to make sure we get a good solid grasp of the foundation (or fundamentals) the coin is based on so you can determine its true value. Once you know the true value, then you can look at the price of the coin to decide if it is inflated or not. Technical analysis, on the other hand, is more focused on studying the actual price movements. A good percentage of the first two chapters of this book was an introduction to the fundamentals of Ethereum, but a fundamental analysis involves much more than just understanding how the system works.

Fundamental Analysis

Usually, fundamental analysis is used to decide which coin you want to invest in but even if you already have your sites on Ethereum, it still helps to do a fundamental analysis to determine if it is stable enough to support putting your trust in it. Ideally, you want to be able to do a fundamental analysis on any coin you want to invest in. Here are a few things you need to examine.

Price stability: A coin's price is considered stable when the number of buyers and the number of sellers are the same. When you place an order to buy Ether, the price would naturally go up. If you place a sell order, then logically, the price would go down. If you have more buyers than sellers, the price will continue to rise and vice versa with sells. However, if the buyers and sellers are evenly matched, the price will be "stable." If the price is stable, it is actually the best time to put your money in the pot.

How much security is needed to protect the Blockchain: The whole purpose of assigning hashes to a transaction is for security.

Demand: Several factors can impact the demand for Ether. These include transaction activity, the amount of trading happening, and user adoption.

Supply: this information is much easier to figure out than demand. The current supply of the coin is how many are in circulation at the time. You also want to look at the rate of new supply that is being released. Each time a miner adds a new block to the chain, he is paid in a certain amount of Ether, adding more to the supply.

Major events: It is important to follow the news to see what major events are happening or are on the horizon. Because these events often affect the perception of the public, they can impact trade on a number of levels.

Once you have collected all this information, you will be able to create a picture of Ethereum, and a pattern should emerge. This type of data can be crucial in deciding at what point you want to put your money in the pot and when it might be best to hold back a little and wait for things to settle down a bit.

Technical Analysis

Conducting a technical analysis is much more detailed than a fundamental analysis, but it is just as important. In fact, many professional analysts believe that it is the technical analysis that can give you a better feel for the market at any given time. It is the tool that allows you to identify specific trends, and even predict future movements in the market. This skill is crucial for anyone who is interested in getting large returns on their investment dollars.

Technical analysis is about looking at the history of the coin. As you study the different charts and graphs, you will begin to see a consistent pattern of price movements you won't see by watching the price periodically. These movements can help you identify repeat patterns and what to expect in the coming days, weeks, months, and even years.

To perform a good technical analysis, you need to be able to make a few assumptions that serve as a valuable foundation on which to base your conclusions on.

The current price on the market is a compilation of all information relating to the coin.

Price movements are not random. They follow trends and patterns. Some patterns are long term while others may be short term but you are searching for those patterns.

History is more valuable to your technical analysis than price movement.

History often repeats itself, and therefore, price movements are predictable.

How to Identify Trends

The ability to determine the general direction a price is moving can be a valuable tool for any type of investment. Still, learning this skill can be quite difficult, especially in a volatile market like cryptocurrency.

However, by applying the assumptions listed above, you know that you can look beyond the present day movements and identify when a coin is about to make an upward move or start heading the other way. There will also be times when the price stays pretty much where it is with little to no upward or downward movement. These are called sideways trends. After studying the charts, you will notice that not only are there the up, down, and sideways trends but each trend has a time frame you must factor in. You might be looking at a short-term uptrend or a long-term sideways trend.

So, what are the tricks that will help you to identify these trends and what should you be looking for?

Moving averages: These are patterns that begin to average out the price of the coin so you can see exactly what it is selling for. By studying the moving average charts, you remove all the price fluctuations, which will make it much easier to see the true direction of a price movement.

There are several different types of moving average charts depending on the length of time you want to analyze. To get a moving average, simply calculate the average price over the specified time period you want to see. Depending on what your investment goal is you could do an average of as little as five days to as many as five years.

There is another moving average you might also want to study, the exponential moving average. It is a moving average that gives more weight to the most recent data. You will consult this average when you want to predict price changes as a result of more recent activity.

These are just a few of the strategies that you might want to think about when getting ready to buy Ethereum. When you understand the patterns that dictate the price movements, you will not only be able to enter the market with more confidence, but you'll also be able to better determine the right time and the right price to get started.

Where to Look for More Information

Knowing where to look to find the information to analyze is not always obvious. Go online, and you'll find hundreds of websites offering you all types of analysis, but you're never quite sure if they can be trusted or not. Cryptocurrency is an entirely new industry, and so are the many people who claim to be experts. It helps to know exactly where to go to find reliable information that you can base your assumptions, predictions, and decisions on.

The White Paper: One of the first places you want to go is the white paper. This is a detailed proposal created by the development team, which details the actual purpose of the coin as well as the mechanics. You won't need this information when doing a technical analysis, but it is very useful for your fundamental approach.

Slack Channel or Blog: Sometimes the white paper can be very technical and is full of complicated jargon that you may not understand. When you have questions, you can go to the slack channel or the coin's blog. This is where the development team interacts with the community and is the perfect platform for clarification when you need it.

Reddit – Steemit: These are public forums where you can feel out how the community actually feels about the coin. You might be surprised at the things you will learn when you're on these sites. It is one of the best ways to determine if there is enough support for the coin or if there is a growing sense of doubt about its future. As you read through these sites, you may hear questions raised that you may never have thought about or you may get better clarification on something that you were still trying to grasp.

It can be a challenge to learn how to analyze any cryptocurrency. There is a lot of information to sift through, and if you're not a detail-oriented person, it may begin to feel overwhelming. However, as you grow in your knowledge of these details, it will eventually become second nature to you, and you'll be much more comfortable in making your predictions.

Chapter 14: How to Invest in Ethereum & Cryptocurrencies

Ethereum

One of the most popular cryptocurrencies in the market today is Ethereum. It is only among a handful of cryptocurrencies attracting serious attention from investors around the world. Since March 2016 to date, Ethereum has gained over 2700% in value. This is the fastest and most impressive growth of any cryptocurrency ever.

Ethereum continues to attract the interest of investors from all over the world. The alternative cryptocurrency, Bitcoin, has been very volatile in recent months. Sharp drop in prices has been driven by regulations instituted by the Chinese government. Regulators in China have put restrictions on ICOs or initial coin offerings which are financial appeals to investors to inject funds into virtual currencies.

Ethereum is quite different from Bitcoin. It is more than just a cryptocurrency but a protocol with plenty of potential. Economists are of the view that its price will be pegged more on its technology than the entire cryptocurrency market. Plenty of experts are expecting the price of Ethereum to rise even further in the last quarter of 2017.

Investing in cryptocurrencies

Virtual currencies such as Ethereum and Bitcoin are some of the most popular investment products in the market today. These digital currencies have the potential of becoming global currencies in the coming years. They may become the preferred payment mode used in international trade, travel, and tourism.

There are different ways of investing in these currencies. Some investors prefer to invest in a couple of cryptocurrencies in additional to investments in other securities. Others are of the view buying and holding major cryptocurrencies means having a share in these important ventures. If an investors puts their funds in

Bitcoin and Ethereum, their value, over time, will far exceed $10,000.

Buy and hold

One of the best investment approaches with cryptocurrencies is to buy and hold. If the value of Ethereum has risen by over 2700% in just 2 years, it is likely to rise phenomenally in the next 5 years. The same is expected to happen with Bitcoin and other important cryptocurrencies.

Caution: Investors are advised to only invest money which they can afford to lose. That is, should the investment lose money, the investor should still be able to keep on living.

The importance of investing in cryptocurrencies

There are sound reasons for any person to invest in cryptocurrencies. The first reason is hedging against the fall or eventual collapse of the dollar. There are those who believe the dollar may eventually fall.

Cryptocurrencies have great attributes not found in other currencies. For instance, they are not regulated by a government or other entity, no one can shut down a user's account or deny them a chance to open one, among many others.

Many investors love the technology behind cryptocurrencies and appreciate that virtual currencies are for every person around the world.

Potential returns from cryptocurrency investment have been extremely attractive and most investors have enjoyed a phenomenal return on investment. This trend is expected to continue for many decades as more and more people around the world turn to virtual currencies.

Where to start

To invest in cryptocurrencies, an investor has to purchase their preferred digital currency at an exchange. There are a couple of reputable exchanges on the

Internet where anyone can purchase Ethereum and Bitcoin and altcoins.

Examples of popular cryptocurrency exchanges and platforms include GDAX, Coinbase and Bitfinex. Others include LocalBitcoins, CoinSource and SatoshiPoint. All these online platforms offer cryptocurrencies and anyone can open an account and purchase the cryptocurrency of their choice.

At the start, prior to 2016, there was only one reputable or well known cryptocurrency. Investors did not have much choice and couldn't choose from among a host of other digital currencies. This is different today because aside from Bitcoin, there is a host of other cryptocurrencies. They include Ethereum, Litecoin, Dash, Ripple and many more. However, most people still trust and have faith in Bitcoin.

Digital wallets

Before buying any cryptocurrency, buyers and investors should acquire an appropriate wallet. Digital wallets are designed to securely store cryptocurrencies, keeping them safe and away from hackers and thieves. It is very important to acquire a secure cryptocurrency wallet before investing.

For a long time, there was a notion that digital wallets hold cryptocurrencies. This is actually not accurate. Any virtual currencies that is purchased remains on the blockchain. What is issued or awarded are the keys. A buyer receives two distinct keys once they purchase digital currencies. These are a private and public key. The private key is the most important and appears as a string of characters. This key allows a buyer access to their digital currencies held within the blockchain.

Buying cryptocurrencies

Now, with a digital wallet at hand, an investor can proceed to their platform or exchange of choice and purchase the currency that they need. There are a couple of considerations to make when choosing a platform. For instance, how reputable is the platform, how long has it been in existence, are they genuine? All these are genuine questions that should be asked. Fortunately, the exchanges mentioned above are all reputable and trustworthy.

Another consideration to make is the cost of the purchase. Different platforms and exchanges charge buyers differently. For instance, Coinbase, which is the most popular platform with Americans, charges a flat fee of 0.5% and an additional 2 – 4% per trade. Platforms like GDAX are a lot more affordable and their charges are pretty low. The platform that a buyer selects will determine the costs, fees and charges that they will incur.

App-based digital wallets

One of the easiest ways of accessing and buying cryptocurrencies is through mobile based digital wallets. These are apps that can be used to conveniently acquire cryptocurrencies.

A good example of an app-based wallet is bread-Wallet. This app is available for both Android and iOS. It is designed to be simple to use, secure and reliable. A tiny amount of the value of cryptocurrency earned is often sent to the miners so they confirm the digital wallet and add it to the blockchain.

The bread-wallet stores the keys on the phone and removes them online. This is a very secure way of storing cryptocurrencies. However, if the phone gets lost or the data gets compromised, the digital currencies could disappear forever. This is why it is essential to back up the information in the wallet and then take the currencies offline. This is a sure way of keeping them safe.

An important point to note is that not all wallets work with all cryptocurrencies. Some are specific, for instance bread-wallet only works with Bitcoin. Therefore, buyers need to confirm the capacity of their wallets before acquiring one. Other wallets though, such as Jaxx, accept multiple currencies, including Ethereum and Bitcoin. Buyers should always confirm this fact to avoid delays and disappointments.

Storing cryptocurrencies with third parties

Those who wish not to have their own digital wallets can store their cryptocurrencies with third parties such as Coinbase. Coinbase is the most trusted cryptocurrency platform. They securely store cryptocurrencies on behalf of their clients. The platform is very stable and has managed to raise more than $117 million from investors such as Horowitz and the Tokyo Mitsubishi bank.

Funds held on this platform are insured and Coinbase can hold both Ethereum and Bitcoin. They can store both public and private keys on behalf of their clients. For security purposes, new investors can choose Coinbase as the place to store their cryptocurrencies.

Point of information

When investing in cryptocurrencies, there are certain features to look out for to determine genuine versus dud currencies. All reputable cryptocurrencies have an active development team, the technology used is very transparent and they have active communities. On the other hand, dud altcoins talk of unclear technical advantages, promote MLM like products, keep insisting they are the next Bitcoin and often target uninformed customers.

It is important to always be on the lookout because there are plenty of new altcoins coming up with others dying off. Plenty of investors have lost money while others have made plenty of profits depending on the path they chose when investing.

Buying cryptocurrencies directly

It is possible to avoid all the high fees and charges when buying cryptocurrencies. This is through direct buying. Buyers are able to buy altcoins and bitcoins directly from sellers. This mostly happens on cryptocurrency exchanges. These exchanges are now found all across Europe, Asia and America.

In the US we have BitFinex, BitStamp and Gemini while in Europe we have Kraken and Bitcoin.de. In Asia, the crypto exchanges include OKCoin, BTCChina and BitFlyer.

Purchasing digital currencies

All these exchanges require that a customer opens an account using their personal information such as names and email. This is required by law due to money laundering regulations. Once the account is open, it is funded in various ways such as with a credit card or direct bank transfer. The accounts are funded using ordinary or fiat money such as the dollar or euro.

As an investor, it is advisable to identify and select an exchange as close as possible, if possible where jurisdiction applies so that investors can recover their funds should anything go wrong. Otherwise, it is advisable to use exchanges in stable countries like Europe where the legal systems can be relied upon.

Buyers and investors enjoy better prices at smaller exchanges with some patience. Larger exchanges can provide large amounts of cryptocurrencies but at high prices and much faster. Smaller amounts at better prices can be purchased at the smaller exchanges so it is better to decide how an investor wishes to buy their cryptocurrencies. Patience always pays so those who buy and hold will benefit more.

Apart from the regular exchanges where buyers and sellers congregate, there are platforms known as altcoin supermarkets that sell all other types of altcoins. Remember altcoins are all cryptocurrencies apart from Bitcoin. Common altcoin exchanges include Yunbi, Bittrex, and Bithumb among others. There are also sites that provide information useful to buyers such as listing of all crypto exchanges and so on.

When is the best time to buy?

In the cryptocurrency world, there is no rule that determines the best time to buy digital coins. However, experts caution buyers to never buy when a currency is crashing or at the peak of a bubble. It is advisable to buy when a currency and the markets are stable and when the price is low.

Trading cryptocurrencies basically depends on two factors. These are when a currency is in a bubble that's about to burst and when it is about to bottom out after crashing. Such points are difficult to determine so it is advisable for those seeking short term gain to purchase when the trend is downwards and sell when the price is on an upward trend.

When investing in multiple currencies, it is best to store them on an exchange. Otherwise it will be too tasking to install malware for each, sync, update and install and much more. In Europe, a good platform to buy cryptocurrencies is Bitcoin.de. However when investing in large amounts of cryptocurrencies, it is not wise to leave them lying on any exchange. This is considered too risky. It is much better to take the risk and store them offline. This is a risk that buyers should take.

Storing digital currencies

It is really advisable to personally store digital currencies. It is possible to do so with proper software but without the need of assistance or trusting another person. There are free and open software for safely storing cryptocurrencies. However, all the data should be backed up and proper malware should be used. This is because should anything go wrong such as a computer crash or malware attack. There are plenty of alternatives regarding secure storage of cryptocurrencies. When proper measures are taken to secure the digital currencies, then they will remain safe, away from hackers and thieves.

Investing in Ethereum

Apart from investing in the most popular cryptocurrency, Bitcoin, many investors are now focusing on the second most popular one which Ethereum. Ethereum is essentially a blockchain-based technology that is more of a protocol than a currency. However, it is largely considered a currency because it uses tokens known as ethers, represented as ETH in the world of digital currencies.

Ethereum's popularity continues to grow across America and around the world. Investors are putting their funds in this crypto because of its excellent performance, its future potential and its unique structure which is different from that of Bitcoin.

Ethereum currently has a market capitalization worth $27 billion which is roughly equal to 40% that of Bitcoin. This is quite impressive considering the phenomenal growth of Bitcoin as well as the death or poor performance of other altcoins. At its best performance, Ethereum was valued at $400 while it is currently $300 as of October 2017. Most investors peg its success to that of Bitcoin and it has performed just as well as Bitcoin. This is why it is often referred to as Bitcoin's little brother.

Ethereum is viewed more like diamond in the world of cryptocurrencies. This is because it has both intrinsic value and underlying value. Bitcoin is a lot more like gold. It is widely traded for its value but has no industrial or underlying value.

Digital wallet

The first step is to acquire a digital wallet to store the cryptocurrency. Not all wallets can store Ethereum so the recommended once are on Coinbase. Coinbase is free and has bonus offers and an app for Smartphone users. The wallet will store ethers which are the actual units of Ethereum.

Where to buy Ethereum (ether)

Investors seeking to buy or invest in Ethereum actually buy ethers. Ether is the currency used on the Ethereum platform.

There are online brokerage firms and digital currency platforms where ethers can be purchased. Like previously explained earlier, it is important to conduct due diligence before buying. For instance, the reliability of vendors should be determined.

Coinbase is a trustworthy platform where different cryptocurrencies can be purchased. These include Ethereum, Bitcoin, and Litecoin. The site offers free wallets to store crypto coins so it is a huge advantage to use this platform.

Important points to note

It is important for buyers to note that when buying ethers, they are purchasing a currency and not investing in an asset. Ethers are not a kind of stock and there are no dividends, profits or payouts. Investing in Ethereum is only in the hope that in the future other investors will gladly purchase the same ethers but at a higher price.

Coinbase is a great place to buy ethers and other cryptocurrencies. For instance, once a buyer opens an account at Coinbase, they can fund it and then buy the ethers on this site and even store them here. The platform provides a digital wallet for free and members even get bonuses when they buy $100 worth of digital currency.

It is a good idea to invest small and make small additional purchases of Ethereum with time. Investors or buyers are advised against make large purchases because a dip in price of Ethereum could see them incur losses. Also, investors should only put in money which they can afford to lose, just in case the currency collapses.

Also, it is possible to set up the account such that automatic Ethereum purchases are made regularly. This works well so that each work or every month, ethers worth $100 or $50 or $1000 are purchased depending on settings and preferences. The only thing to note is that price of Ethereum varies month on month. Therefore, different ether amounts will be obtained each time even when bought by the same quantity of cash.

Buy and hold strategy

When an investor buys Ethereum, one of the best strategies is to hold onto it for a while, preferably a year. There are often huge price changes within that space of time such that buyers get significant returns on their investments.

For instance, in the last 2 years alone, Ethereum has gained over 2700% in value from January 2016 to October 2017. Holding onto it for a reasonable period of time is likely to see positive yields. Patience is key here. Long term investment in cryptocurrencies and especially in products such as Ethereum is the best way of creating wealth and hedging positively.

Avoid day trading at all costs

Investors may be tempted to engage in day trading in order to make quick profits. Day trading is the process of buying small and frequently during the day when prices are low and then sell when the price increases. This is a highly speculative trade and very, very risky. More often than not, investors lose money. Even the most seasoned traders experience challenges engaging in day trading. The best form of investment with Ethereum is to buy and hold. It is also advisable to buy small quantities and keep increasing investment periodically.

Sometimes buyers seek to profit at the slightest change in price. In fact, many investors are tempted to sell at the slightest change in price. This is not good practice because many others will rush and do the same, causing a deep fall in value and plenty of volatility in the market.

It is always the responsibility of a buyer to keep their Ethereum safe. The benefit of storing cryptocurrencies at Coinbase is that they are totally secure as the sight is tamperproof and all currencies are insured.

Conclusion

 let's hope it was informative and able to provide you with all of the tools you need to achieve your goals, whatever it is that they may be. Just because you've finished this book doesn't mean there is nothing left to learn on the topic, expanding your horizons is the only way to find the mastery you seek. Ethereum is still an extremely new technology which means that new chapters in its story are always emerging. As such, if you hope to take advantage of its growth for you own ends then you are going to need to become a dedicated lifelong learner as if you stop, you never know what you might miss and what it might cost you.

For now, however, as you have finished this book, it is time to start focusing on how you are going to take advantage of all of the possibilities that Ethereum has to offer. There is so much going on in the space right now, that it doesn't even really matter what you choose, be it mining, trading or programming, as long as you get out there and really choose one of them. Cryptocurrency is the Next Big Thing and Ethereum is already poised to be on the top when things really take off. Don't look this gift horse in the mouth, pick something and get in on the ground floor of it or in five years you will wish you had.

9 781990 373640